Barron's Regents Exams and Answers

Math A

D0068268

LAWRENCE S. LEFF
Assistant Principal
Mathematics Supervisor
Franklin D. Roosevelt High School, Brooklyn, NY

BARRON'S

Barron's Educational Series, Inc.

All inquiries should be addressed to:
Barron's Educational Series, Inc.
250 Wireless Boulevard
Hauppauge, NY 11788
http://www.barronseduc.com

ISBN 0-7641-1552-9
ISSN 1528-1036

PRINTED IN THE UNITED STATES OF AMERICA
9 8 7 6 5 4 3

Contents

1. Know What to Expect on Test Day
2. Avoid Last-Minute Studying
3. Be Well Rested and Come Prepared on Test Day
4. Know How to Use Your Calculator
5. Know When to Use Your Calculator
6. Have a Plan for Budgeting Your Time
7. Make Your Answers Easy to Read
8. Answer the Question That Is Asked
9. Take Advantage of Multiple-Choice Questions
10. Don't Omit Any Questions

Special Problem-Solving Strategies 20

1. Work Backwards
2. Make a Table or List
3. Draw a Diagram
4. Use Particular Numbers
5. Guess and Check

A Brief Review of Key Math A Facts and Skills 26

1. Number Concepts
2. Logic
3. Algebraic Methods
4. Geometry
5. Measurement
6. Symmetry, Transformations, and Graphing
7. Systems of Equations and Inequalities
8. Probability and Statistics

Glossary of Terms 149

Regents Examinations, Solutions, Answers, and Self-Analysis Charts 161

Preface

This book is designed to prepare you for the Math A Regents examination while strengthening your understanding and mastery of the material on which this test is based.

In addition to providing complete sets of questions from previous Mathematics A Regents examinations, this book offers these special features:

- **Step-by-Step Explanations of the Solutions to All Regents Questions**. Careful study of the solutions and explanations will improve your mastery of the subject. Each explanation is designed to show you how to apply the facts and concepts you have learned in class. Since the explanation for each solution has been written with emphasis on the reasoning behind each step, its value goes well beyond the application to a particular question.

- **Unique System of Self-Analysis Charts**. Each set of solutions for a particular Math A Regents exam ends with a Self-Analysis Chart. These charts will help you to identify weaknesses and direct your study efforts where needed. In addition, the charts classify the questions on each exam into an organized set of topic groups. This feature will enable you to locate other questions on the same topic in other Math A Regents exams.

- ***General Test-Taking Tips***. Tips are given that will help to raise your grade on the actual Math A Regents exam that you will take.

- ***Special Mathematics Problem-Solving Strategies***. Problem-solving strategies that you can try if you get stuck on a problem are explained and illustrated.

- ***Mathematics A Refresher***. A brief review of key mathematics facts and skills that are tested on the Math A Regents exam is included for easy reference and quick study.

- ***Glossary***. Definitions of important terms related to the Math A Regents exam are conveniently organized in a glossary.

Frequency of Topics— Math A

Questions in the Math A Regents exams fall into one of 31 topic categories. The "Brief Review of Key Math A Facts and Skills" that appears later in this book covers most of these concepts.

The Frequency Chart shows how many questions in recent exams have been in each category, to indicate which topics have been emphasized in recent years.

The Self-Analysis Charts that follow each Regents exam designate exactly which questions in that exam are in each category, so you can determine where your weakest areas are. You may also try questions in those areas again, for more practice.

The two charts—Frequency and Self-Analysis—should give you a very good idea of the topics you need to review, and the Practice Exercises in the "Brief Review" provide more practice.

Frequency of Topics—Math A

Number of Questions

	Sample	June 1999	Aug 1999	Jan 2000	June 2000	Aug 2000	Jan 2001	June 2001
1. Numbers; Properties of Real Numbers; Percent	1	2	2	3	1	1	1	3
2. Operations on Rational Numbers & Monomials	—	—	1	2	1	1	—	—
3. Laws of Exponents for Integer Exponents; Scientific Notation	1	1	1	1	1	1	1	—
4. Operations on Polynomials	2	1	—	1	1	1	1	1
5. Square Root; Operations with Radicals	1	1	1	1	—	1	1	—
6. Evaluating Formulas & Alg. Expressions	—	1	1	1	1	—	1	—
7. Solving Linear Eqs. & Inequalities	2	—	1	—	1	1	—	1
8. Solving Literal Eqs. & Formulas for a Particular Letter	—	1	—	1	1	—	1	—
9. Alg. Operations (including factoring)	1	2	1	1	1	2	2	1
10. Solving Quadratic Eqs.	—	1	2	—	2	2	2	1
11. Coordinate Geometry (graphs of linear eqs.; slope; midpoint; distance)	1	2	2	2	1	2	1	1
12. Systems of Linear Eqs. & Inequalities (alg. and graph. solutions)	1	—	—	—	1	1	—	1
13. Mathematical Modeling Using: Eqs., Tables, Graphs of Linear Eqs., Parabolas	1	1	1	2	1	1	1	2
14. Linear-Quad Systems (alg. and graphical solutions)	—	1	—	1	1	—	—	1
15. Word Problems Requiring Arith. or Algebraic Reasoning	5	4	5	3	3	4	7	6

		Sample	June 1999	Aug 1999	Jan 2000	June 2000	Aug 2000	Jan 2001	June 2001
						Number of Questions			
16.	Areas, Perims., Circumf., Vols. of Common Figures	2	4	6	2	1	5	1	4
17.	Angle & Line Relationships (suppl., compl., vertical angles; parallel lines; congruence)	2	—	—	—	—	—	1	1
18.	Ratio & Proportion (incl. similar polygons)	—	1	—	2	2	1	—	1
19.	Pythagorean Theorem	1	—	—	1	1	—	1	1
20.	Right Triangle Trig. & Indirect Measurement	1	1	1	—	1	1	1	—
21.	Logic (symbolic rep.; conditionals; logically equiv. statements; valid arguments)	1	1	1	1	1	2	2	3
22.	Probability (incl. tree diagrams & sample spaces)	2	2	1	1	—	1	2	1
23.	Counting Methods & Sets	1	1	1	—	1	1	1	—
24.	Permutations & Combinations	—	1	1	2	3	1	—	2
25.	Statistics (mean, percentiles, quartiles; freq. dist.; histograms; stem & leaf plots)	3	2	2	3	2	1	2	—
26.	Properties of Triangles & Parallelograms	1	3	2	2	1	1	2	3
27.	Transformations (reflections; translations; rotations; dilations)	1	1	1	1	1	—	—	—
28.	Symmetry	—	1	—	—	1	—	1	—
29.	Area and Transformations Using Coordinates	2	—	—	—	—	1	1	1
30.	Locus & Constructions	1	—	1	1	2	1	1	—
31.	Dimensional Analysis	—	—	—	—	1	—	—	—

How to Use This Book

This section explains how you can make the best use of your study time.

As you work your way through this book, you will be following a carefully designed five-step study plan that will improve your understanding of the topics that the Math A Regents examination tests while raising your exam grade.

Step 1. *Know What to Expect on Test Day.* Before the day of the test, you should be thoroughly familiar with the format, the scoring, and the special directions for the Math A Regents exam. This knowledge will help you build confidence and prevent errors that may arise from misunderstanding the directions. The next section in this book, "Getting Acquainted with Math A," provides this important information.

Step 2. *Become Testwise.* The section titled "Ten Test-Taking Tips" will alert you to easy things that you can do to become better prepared and to be more confident when you take the test.

Step 3. *Learn Problem-Solving Strategies.* The strategies explained in the section entitled "Special Problem-Solving Strategies" may help you solve problems that at first glance seem too difficult or complicated.

Step 4. Review Mathematics A Topics. Since the Math A Regents exam will test you on topics that you may not have studied recently, the section entitled "A Brief Review of Key Math A Skills" provides a quick refresher of the major topics tested on the examination. This section also includes illustrative practice exercises with worked-out solutions.

Step 5. Take Practice Exams Under Exam Conditions. The greater part of this book contains actual Math A Regents exams with carefully worked out solutions for all of the questions. When you reach this part of the book, you should do these things:

- After you complete an exam, check the answer key for the entire test. Circle any omitted questions or questions that you answered incorrectly. Study the explained solutions for these questions.

- On the Self-Analysis Chart, find the topic under which each question is classified, and enter the number of points you earned if you answered the question correctly.

- Figure out your percentage on each topic by dividing your earned points by the total number of points allotted to that topic, carrying the division to two decimal places. If you are not satisfied with your percentage on any topic, reread the explained solutions for the questions you missed. Then locate related questions in other Regents examinations by using their Self-Analysis Charts to see which questions are listed for the troublesome topic. Attempting to solve these questions and then studying their solutions will provide you with additional preparation. You may also find it helpful to review the appropriate sections in "A Brief Review of Key Math A Skills." More detailed explanations and additional practice problems with answers are available in Barron's companion book, *Let's Review: Mathematics A.*

Tips for Practicing Effectively and Efficiently

• In taking a practice test, do not spend too much time on any one question. If you cannot come up with a method to use, or if you cannot complete the solution, put a slash through the number of the question. When you have completed as many questions as you can, return to the unanswered questions and try them again.

• When you have finished the practice test, compare each of your solutions with the solutions that are given. Read the explanation provided even if you have answered the question correctly. Each solution has been carefully designed to provide additional insight that may be valuable when answering a more difficult question on the same topic.

• In the weeks before the actual test, plan to devote at least one-half hour each day to preparation. It is better to spread out your time in this way than to cram by preparing for, say, three hours in one evening. As the test day gets closer, take at least one complete Regents exam under actual test conditions.

Getting Acquainted with Math A

This section explains things about the Math A Regents examination that you may not know, such as how the exam is organized, how your exam will be scored, and where you can find a complete listing of the topics tested by this exam.

WHAT IS MATH A?

The Math A examination was offered for the first time in June 1999. It will eventually replace the Regents Sequential Math Courses I and II. January 2002 is the last opportunity to take Sequential Math Course I. If you are entering high school in September 2001, you must pass either the Math A or Math B Regents examination before you can graduate. Math B was introduced in June 2001.

WHEN DO I TAKE THE MATH A REGENTS EXAM?

The Math A Regents exam is administered in January, June, and August of every school year. Most students will take this exam 1 to 2 years after they begin their study of high-school-level mathematics. Your school and the mathematics program in which you are enrolled will determine when you should take the Math A Regents exam.

HOW IS THE MATH A REGENTS EXAM SET UP?

The Math A Regents exam is divided into four parts with a total of 35 questions. All of the questions in each of the four parts must be answered. You will be allowed a maximum of 3 hours in which to complete the test.

Part I consists of 20 standard multiple-choice questions, each with four answer choices labeled (1), (2), (3), and (4). After you figure out the answer to each multiple-choice question, you must write the numeral that precedes the correct choice in the space provided on the separate tear-off answer sheet for Part I, which is the last page of the question booklet.

The answers and the accompanying work for the questions in Parts II, III, and IV must be written directly in the question booklet. You must show or explain how you arrived at each answer by indicating the necessary steps, including appropriate formula substitutions, diagrams, graphs, and charts. If you use a guess-and-check strategy to arrive at an answer for a problem, you must show at least one guess that does *not* work.

Since scrap paper is not permitted for any part of the exam, you may use the blank spaces in the question booklet as scrap paper. If you need to draw a graph, graph paper will be provided in the question booklet. All work should be done in pen, except graphs and diagrams, which should be drawn in pencil.

WHAT TYPE OF CALCULATOR DO I NEED?

Scientific calculators are *required* for the Math A Regents examination. You will need to use your scientific calculator to work with trigonometric functions of angles, as well as to do routine calculations. A scientific calculator will also be helpful when you evaluate permutations ($_nP_r$) and combinations ($_nC_r$).

Graphing calculators are permitted but *not* required for the Math A examination. Here are three possible advantages of having a graphing calculator and knowing how to use it:

- A graphing calculator allows you to quickly visualize a problem that can be represented by an equation or an inequality.
- Some problems can be solved more easily graphically with a calculator than algebraically with pen and paper.
- A graphical solution with a calculator may help to confirm an answer obtained using standard algebraic methods.

You can find helpful information about how specific types of problems can be solved using a graphing calculator in Barron's companion review book, *Let's Review: Mathematics A*.

WHAT GETS COLLECTED AT THE END OF THE EXAMINATION?

At the end of examination, you must return:

- Any tool provided to you by your school, such as a calculator, straightedge (ruler), or compass.
- The question booklet. Check that you have printed your name and the name of your school in the appropriate boxes near the top of the first page.
- The Part I answer sheet. You must sign the statement at the bottom of the Part I answer sheet indicating that you did not receive any unlawful assistance in answering any of the questions. If you fail to sign this declaration, your answer paper will not be accepted.

HOW IS THE EXAM SCORED?

Your answers to the 20 multiple-choice questions in Part I are scored as either correct or incorrect. Each correct answer receives 2 points.

The five questions in Part II are worth 2 points each, the five questions in Part III are worth 3 points each, and the five questions in Part IV are worth 4 points each. Solutions to questions in Parts II, III, and IV that are not completely correct may receive partial credit according to a special scoring guide provided by the New York State Education Department.

The accompanying table shows how the Math A exam breaks down.

Part	Number of Questions	Point Value	Total Points
I	20 multiple-choice	2 each	$20 \times 2 = 40$
II	5	2 each	$5 \times 2 = 10$
III	5	3 each	$5 \times 3 = 15$
IV	5	4 each	$5 \times 4 = 20$
	Test = 35 questions		Test = 85 points

HOW IS YOUR FINAL SCORE DETERMINED?

The maximum total raw score for the Math A Regents examination is 85 points. After the raw scores for the four parts of the test are added together, a conversion table provided by the New York State Education Department is used to convert your raw score into a final test score that falls within the usual 0 to 100 scale.

IS THE MATH A REGENTS OFFERED IN DIFFERENT LANGUAGES?

In addition to English, the Math A Regents examination is offered in these languages: Chinese, Haitian Creole, Korean, Russian, and Spanish. These alternative language versions of the exam must be ordered by your school well in advance of the day of the exam. If you think it would be to your advantage to take an alternative language version of the Math A Regents exam, you should discuss the matter further with your teacher.

WHAT IS THE *CORE CURRICULUM*?

The *Core Curriculum* is the official publication by the New York State Education Department that lists the topics that may be tested by the Math A Regents examination. The *Core Curriculum* includes most of the topics from Course I of Sequential Mathematics and a selection of topics from Course II. Therefore the Math A Regents exam can test you on a wide range of topics that include:

- Facts about numbers; basic algebraic operations; polynomial arithmetic; factoring.
- Solving linear equations and factorable quadratic equations; solving systems of linear equations and linear-quadratic pairs both algebraically and graphically; solving systems of linear inequalities graphically.
- Logical analysis and basic geometric concepts (similarity and congruence; properties of triangles, parallel lines, parallelograms, and special quadrilaterals; the Pythagorean theorem).
- Measurement; perimeter, area, and volume; trigonometry of the right triangle.
- Transformations; locus; geometric constructions using a compass and straightedge.
- Coordinate geometry; linear equations and inequalities; circles and parabolas.
- Data analysis (circle graphs, scatter plots, histograms, stem-and-leaf plots, box-and-whisker plots).
- Probability, counting methods, and statistics; factorials; permutations and combinations.

There are a few topics from Course II that you do *not* need to study. The Mathematics A Regents examination does not include formal proofs of any kind, so you do not need to know:

- Formal geometric proofs.
- Formal logic proofs.
- Coordinate geometry proofs.
- The quadratic formula.

If you have Internet access, you can view the *Core Curriculum* at the New York State Education Department's web site at
http://www.nysed.gov/rscs/pubs.html#res

Ten Test-Taking Tips

1. Know What to Expect on Test Day
2. Avoid Last-Minute Studying
3. Be Well Rested and Come Prepared on Test Day
4. Know How to Use Your Calculator
5. Know When to Use Your Calculator
6. Have a Plan for Budgeting Your Time
7. Make Your Answers Easy to Read
8. Answer the Question That Is Asked
9. Take Advantage of Multiple-Choice Questions
10. Don't Omit Any Questions

These ten practical tips can help you raise your grade on the Math A Regents examination.

TIP 1

Know What to Expect on Test Day

SUGGESTIONS
• Become familiar with the format, directions, and content of the Math A Regents exam.

- Know where you should write your answers for the different parts of the test.
- Ask your teacher to show you an actual test booklet of a previously given Math A exam.

TIP 2

Avoid Last-Minute Studying

SUGGESTIONS

- Start your Math A Regents exam preparation early by making a regular practice of (1) taking detailed notes in class and then reviewing your notes when you get home, (2) completing all written homework assignments in a neat and organized way, (3) writing down any questions you have about your homework so that you can ask your teacher about them, and (4) saving your classroom tests for use as an additional source of questions.
- Get a review book early in your preparation so that additional help or explanations, if needed, will be at your fingertips. The recommended review book is Barron's *Let's Review: Math A.* This easy-to-follow book has been designed for fast and effective learning.
- Build your skill and confidence by completing all of the exams in this book and studying the accompanying solutions before the day of the Math A Regents exam. Because each exam takes up to 3 hours to complete, you should begin this process no later than several weeks before the exam is scheduled to be given.
- As the day of the actual exam nears, take the exams in this book under the timed conditions that you will encounter on the actual test. Then compare each of your answers with the explained answers given in this book.
- Use the Self-Analysis Chart at the end of each exam to help pinpoint any weaknesses.

- If you do not feel confident in a particular area, study the corresponding topic in Barron's *Let's Review: Math A.*
- As you work your way through the exams in this book, make a list of any formulas or rules that you need to know, and learn them well before the day of the exam.

TIP 3

Be Well Rested and Come Prepared on Test Day

SUGGESTIONS

- On the night before the Math A Regents exam, lay out all of the things you must take with you. Check the items against the following list:

  1. Your Regents admission card with the room number of the exam.
  2. Two pens.
  3. Two sharpened pencils with erasers.
  4. A ruler.
  5. A compass.
  6. A scientific calculator.
  7. A watch.

- If your calculator uses batteries, insert fresh ones the night before the exam.
- Eat wisely and go to bed early so you will be alert and well rested when you take the exam.
- Be certain you know when your exam begins. Set your alarm clock to give you plenty of time to eat breakfast and travel to school. Also, tell your parents what time you will need to leave the house in order to get to school on time.
- Arrive at the exam room on time and with confidence that you are well prepared.

TIP 4

Know How to Use Your Calculator

SUGGESTIONS

- Take to the Math A Regents exam the same calculator that you used when you completed the practice exams at home.
- If you wish, take a graphing calculator also, although no question will depend on its use.
- If you are required to use a calculator provided by your school, make sure that you practice with it in advance because not all calculators work in the same way.
- Know how to use your calculator to find the value of the sine, cosine, or tangent of an angle correct to four decimal places.
- Know how to use your calculator to find the number of degrees in an angle when the value of a trigonometric function of that angle is given. For example, if $\sin x = 0.9511$, you should be able to use your scientific calculator to determine that the value of angle x, correct to the nearest degree, is 72°.
- Become an expert on using your calculator to find factorials $(n!)$, permutations $({}_nP_r)$, and combinations $({}_nC_r)$.
- Because it is easy to press the wrong key, first estimate an answer and then compare it to the answer obtained by using your calculator. If the two answers are very different, start over.

TIP 5

Know When to Use Your Calculator

SUGGESTIONS
- Don't expect to have to use your calculator on each question. Most questions will not require a calculator.
- Expect to have to use your calculator when solving numerical problems involving the three trigonometric ratios: sine, cosine, and tangent.
- Get into the habit of using your scientific calculator to evaluate factorials, permutations, and combinations because these calculations are prone to error when performed manually using the appropriate formulas.

TIP 6

Have a Plan for Budgeting Your Time

SUGGESTIONS
- In the **first hour** of the 3-hour Math A Regents exam, complete the 20 multiple-choice questions in Part I. In answering troublesome questions of this type, first rule out any choices that are impossible. If the choices are numbers, you may be able to identify the correct answer by plugging these numbers back into the original question to see which one works. If the choices are letters, you can substitute easy numbers for the letters in both the test question and in each choice and then try to match the numerical result produced in the question to the answer choice that evaluates to the same number. This suggestion is explained more fully in Tip 9.

- In the **second hour** of the exam, complete the five Part II questions and the five Part III questions. To maximize your credit for each question, write down clearly the steps you followed to arrive at each answer. Include any equations, formula substitutions, diagrams, tables, graphs, and so forth.
- In the **last hour** of the exam:
  1. Complete the five Part IV questions. Again, be sure to show how you arrived at each answer.
  2. During the last 10 minutes, review your entire test paper for neatness and accuracy. Check that all answers (except graphs and other drawings) are written in ink. Make sure you have answered all of the questions in each part of the exam and that all of your Part I answers have been recorded accurately on the separate Part I answer sheet.
  3. Before you submit your test materials to the proctor, check that you have written your name in the reserved spaces on the front page of the question booklet and on the Part I answer sheet. Also, don't forget to sign the declaration that appears at the bottom of the Part I answer sheet.

TIP 7

Make Your Answers Easy to Read

SUGGESTIONS

- Make sure your solutions and answers are clear, neat, and logically organized. When solving problems algebraically, define what the variables stand for, as in "Let $x = \ldots$."
- Use a pencil to draw graphs so that you can erase neatly, if necessary. Use a ruler to draw straight lines and axes.
- Do not forget to label the coordinate axes. Put y at the top of the vertical axis and $-y$ at the bottom. Write x to the right of the horizontal axis and $-x$ to the left. Next to each graph, write its equation.

- When answering a question on Parts II, III, or IV, record your reasoning as well as your final answers. Provide enough details to enable someone who doesn't know how you think to understand why and how you moved from one step of the solution to the next. If the teacher who is grading your paper finds it difficult to figure out what you wrote, he or she may simply decide to mark your work as incorrect and to give you little, if any, partial credit.
- Draw a box around your final answer to a Part II, III, or IV question.

TIP 8

Answer the Question That Is Asked

SUGGESTIONS

- Make sure each of your answers is in the form required by the question. For example, if a question asks for an approximation, round off your answer to the required decimal position. If a question asks that you write an answer in lowest terms, make sure that the numerator and denominator of a fractional answer do not have any common factors other than 1 or −1. For example, instead of leaving the answer as $\frac{10}{12}$, write $\frac{5}{6}$; instead of leaving $\frac{x^2 - 1}{x + 1}$, write $x - 1$ since

$$\frac{x^2 - 1}{x + 1} = \frac{(x - 1)\overset{1}{\cancel{(x + 1)}}}{\cancel{x + 1}} = x - 1$$

If a question calls for the answer in simplest form, make sure you simplify a square root radical so that the radicand does not contain any perfect square factors greater than 1. For example, instead of leaving the answer as $\sqrt{18}$, write $3\sqrt{2}$.
- If a question asks for the x-coordinate (or y-coordinate) of a point, do not give both the x- and the y-coordinates.

- After solving a word problem, check the original question to make sure your *final* answer is the quantity that the question asks you to find.
- If units of measurement are given, as in area problems, check that your answer includes the correct units.
- If a question requires a positive root of a quadratic equation, as in geometric problems in which the variable represents a dimension, make sure you reject the negative root.

TIP 9

Take Advantage of Multiple-Choice Questions

SUGGESTIONS

- When the answer choices of a multiple-choice question contain only numbers, try plugging each choice into the original question until you find the one that works.

EXAMPLE
Which ordered pair is the solution set for this system of equations?

$$x + y = 8$$
$$y = x - 3$$

(1) (2.5,5.5)   (3) (4,4)
(2) (4,1)   (4) (5.5,2.5)

Solution 1: If you don't remember how to arrive at the solution by using an algebraic method, you can get the right answer by trying each ordered pair in the system of equations until you find the pair that works for *both* equations.

Choice (1): For (2.5,5.5), let $x = 2.5$ and $y = 5.5$:

$x + y = 8 \Rightarrow 2.5 + 5.5 = 8$ It works, so test the second equation.

$y = x - 3 \Rightarrow 5.5 \neq 2.5 - 3$ It doesn't work, since $2.5 - 3 = 0.5$.

Choice (2): For (4,1), let $x = 4$ and $y = 1$:

$x + y = 8 \Rightarrow 4 + 1 \neq 8$ It doesn't work, so there is no need to test the second equation.

Choice (3): For (4,4), let $x = 4$ and $y = 4$:

$x + y = 8 \Rightarrow 4 + 4 = 8$ It works, so test the second equation.

$y = x - 3 \Rightarrow 4 \neq 4 - 3$ It doesn't work, since $4 - 3 = 1$.

Choice (4): For (5.5,2.5), let $x = 5.5$ and $y = 2.5$:

$x + y = 8 \Rightarrow 5.5 + 2.5 = 8$ It works, so test the second equation.

$y = x - 3 \Rightarrow 2.5 = 5.5 - 3$ Since it works, (5.5,2.5) is the solution.

Hence, the correct choice is **(4)**.

Solution 2: To get the solution algebraically, eliminate y in the first equation by replacing it with $x - 3$; then $x + (x - 3) = 8$.

$$2x - 3 = 8$$
$$2x = 11$$
$$\frac{2x}{2} = \frac{11}{2}$$
$$x = 5.5$$

The correct choice is **(4)**, the only choice for which $x = 5.5$.

• If the answer choices for a multiple-choice question contain only letters, replace the letters with easy numbers in both the question and the answer choices. Then work out the problem using these numbers.

EXAMPLE (August 1999 Math A, Number 11)

Which expression is equal to $\dfrac{a}{x} + \dfrac{b}{2x}$?

(1) $\dfrac{2a + b}{2x}$ (2) $\dfrac{2a + b}{x}$ (3) $\dfrac{a + b}{3x}$ (4) $\dfrac{a + b}{2x}$

Solution 1: Let $a = b = x = 1$. Then

$$\frac{a}{x} + \frac{b}{2x} = \frac{1}{1} + \frac{1}{2(1)} = 1 + \frac{1}{2} = \frac{3}{2}$$

Next, evaluate each of the four answer choices to find the one that is equal to $\frac{3}{2}$. If more than one answer choice evaluates to $\frac{3}{2}$, start over using different values for a, b, and x.

Choice (1): $\dfrac{2a+b}{2x} = \dfrac{2(1)+1}{2(1)} = \dfrac{2+1}{2} = \dfrac{3}{2}.$

Choice (2): $\dfrac{2a+b}{x} = \dfrac{2(1)+1}{1} = \dfrac{3}{1} = 3.$

Choice (3): $\dfrac{a+b}{3x} = \dfrac{1+1}{3(1)} = \dfrac{2}{3}.$

Choice (4): $\dfrac{a+b}{2x} = \dfrac{1+1}{2(1)} = \dfrac{2}{2} = 1.$

The correct choice is **(1)**, the only choice that evaluates to $\dfrac{3}{2}$.

Solution 2: To solve algebraically, change the first fraction into an equivalent fraction that has the LCD as its denominator. Then write the sum of the numerators of the two fractions over the LCD.

$$\frac{a}{x} + \frac{b}{2x} = \left(\frac{2}{2}\right)\frac{a}{x} + \frac{b}{2x}$$

$$= \frac{2a+b}{2x}$$

Hence, the correct choice is **(1)**.

TIP 10
Don't Omit Any Questions

SUGGESTIONS

- Keep in mind that on each of the four parts of the test you must answer all of the questions.

- If you get stuck on a multiple-choice question in Part I, try one of the problem-solving methods discussed in Tip 9. If you still can't figure out the answer, try to eliminate any impossible answers. Then guess from the remaining choices.

- If you get stuck on a question from Parts II, III, or IV, try to maximize your partial credit by writing down any formula, diagram, or mathematics facts that you think might apply. If appropriate, organize and analyze the given information by making a table or a diagram. You can then try to arrive at the correct answer by guessing, checking your guess, and then revising your guess, as needed.

Special Problem-Solving Strategies

No single problem-solving strategy works for all problems. You should have a "toolbox" of strategies from which you can choose when trying to figure out how to solve an unfamiliar type of problem.

If you get stuck trying to solve a problem, use one or more of the following strategies.

STRATEGY 1: WORK BACKWARDS

Reversing the steps that produced an end result can lead to the starting value that you need to find.

EXAMPLE (June 1999 Math A, Number 10)
Linda paid $48 for a jacket that was on sale for 25% of the original price. What was the original price of the jacket?

(1) $60   (2) $72   (3) $96   (4) $192

Solution: Work back from the final price of $48. Since Linda paid $48 for the jacket that was on sale for 25% $\left(=\frac{1}{4}\right)$ of its original price, the original price was $4 \times \$48 = \$192$.

The correct choice is (**4**).

EXAMPLE (June 1999 Math A, Number 25)

Sara's telephone service costs $21 per month plus $0.25 for each local call; long-distance calls are extra. Last month, Sara's bill was $36.64, and it included $6.14 in long-distance charges. How many calls did she make?

Solution: Work back from the final bill of $36.64.

• The part of the bill that does not include any long-distance charges is $36.64 − $6.14 = $30.50.

• The part of the bill that does not include the long-distance charges or the basic service charge is $30.50 − $21 = $9.50.

• Since the bill for local calls is $9.50 and each local call cost $0.25, the number of local calls is $\frac{\$9.50}{\$0.25} = \mathbf{38}$.

STRATEGY 2: MAKE A TABLE OR LIST

Making a table or list can help you to organize the facts of a problem so that the answer becomes easier to find.

EXAMPLE (January 2000 Math A, Number 35)

The Excel Cable Company has a monthly fee of $32.00 and an additional charge of $8.00 for each premium channel. The Best Cable Company has a monthly fee of $26.00 and an additional charge of $10.00 for each premium channel. For what number of premium channels will the total monthly subscription fee for the two cable companies be the same?

Solution: Make a table to compare the fees for each company.

Number of Premium Channels	Fee Charged by the Excel Cable Company	Fee Charged by the Best Cable Company
1	$32 + (1 × $8) = $40	$26 + (1 × $10) = $36
2	$32 + (2 × $8) = $48	$26 + (2 × $10) = $46
3	$32 + (3 × $8) = $56	$26 + (3 × $10) = $56

The fees for the two cable companies are the same for **3** premium channels.

STRATEGY 3: DRAW A DIAGRAM

Drawing a diagram when none is given can help you to visualize the facts of a problem and thereby may lead to a method for solving the problem.

EXAMPLE (Sample Math A Test, Number 29)
In a school of 320 students, 85 students are in the band, 200 students are on sports teams, and 60 students participate in both activities. How many students are *not* involved in either band or sports? Show how you arrived at your answer.

Solution: Draw a diagram like the one below.

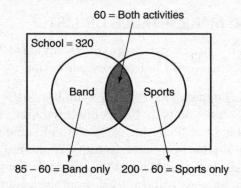

60 = Both activities

School = 320

Band   Sports

85 – 60 = Band only   200 – 60 = Sports only

- 60 students are involved in both activities.

- 85 – 60 = 25 students are involved only in the band.

- 200 – 60 = 140 students are involved only in sports.

- Therefore, 320 – 60 – 25 – 140 = **95** students are *not* involved in either of the two activities.

STRATEGY 4: USE PARTICULAR NUMBERS

Using numbers instead of variables can greatly simplify a problem.

EXAMPLE (August 1999 Math A, Number 18)
The ratio of the lengths of corresponding sides of two similar squares is 1 to 3. What is the ratio of the area of the smaller square to the area of the larger square?

(1) $1:\sqrt{3}$ (2) 1:3   (3) 1:6   (4) 1:9

Solution: Since it is given that the ratio of lengths of corresponding sides of two similar squares is 1 to 3, let the length of a side of the smaller square equal 1 and the length of the corresponding side of the larger square equal 3. Then:

- The area of the smaller square is $1 \times 1 = 1$.
- The area of the larger square is $3 \times 3 = 9$.
- The ratio of the area of the smaller square to the area of the larger square is 1:9.

Hence, the correct choice is **(4)**.

In a multiple-choice question, changing variable answer choices into numerical choices, as illustrated in Tip 9 on pages 16 to 18, can make the question easier to answer.

STRATEGY 5: GUESS AND CHECK

Even if you can't solve an unfamiliar problem, you may have enough information to make a reasonable guess.

EXAMPLE (June 1999 Math A, Number 15)
During a recent winter, the ratio of deer to foxes was 7 to 3 in one county of New York State. If there were 210 foxes in the county, what was the number of deer?

(1) 90   (2) 147   (3) 280   (4) 490

Solution: Make a reasonable guess, and then check it against the four answer choices.

- Because the ratio of deer to foxes was 7 to 3, the number of deer was a little more than twice the number of foxes.
- Since it is given that there were 210 foxes in the county, the number of deer must be a little more than $2 \times 210 = 420$.
- Of the four answer choices, only 490 is greater than 420.

Hence, the correct choice is **(4)**.

EXAMPLE

Farmer Gray has only chickens and cows in his barnyard. If these animals have a total of 60 heads and 140 legs, how many chickens and how many cows are in the barnyard?

Solution 1:
- Since there are 60 heads, the sum of the number of cows and the number of chickens is 60. A chicken has two legs and a cow has four legs.
- Guess, check, and revise while keeping track of your guesses in a list or a table.

Number of Chickens	Cows	Total Number of Legs
0	60	$(0 \times 2) + (60 \times 4) = 240$ ← Too high
30	30	$(30 \times 2) + (30 \times 4) = 180$ ← Too high
40	20	$(40 \times 2) + (20 \times 4) = 160$ ← Too high
50	**10**	$(50 \times 2) + (10 \times 4) = 140$ ← This is the answer!

- Check that your answer works. Two conditions must be true:

✓ The total number of heads is 60:   50 chickens + 10 cows = 60 heads

✓ The total number of legs is 140:   50 chickens × 2 legs   = 100 legs

10 cows × 4 legs   = 40 legs
Total   = 140 legs

Farmer Gray has **50 chickens** and **10 cows**.

Solution 2:   Let x = number of chickens.
Then $60 - x$ = number of cows. Hence:

$$2x + 4(60 - x) = 140$$
$$2x + 240 - 4x = 140$$
$$-2x + 240 = 140$$
$$-2x = -100$$
$$x = \frac{-100}{-2} = 50$$

Since there are 50 chickens, there are $60 - 50 = 10$ cows.

Farmer Gray has **50 chickens** and **10 cows**.

A General Problem-Solving Approach

1. Read **_each test question through the first time_** to get a *general idea* of the type of mathematics knowledge that is required. For example, one question may ask you to solve an algebraic equation, while for another question you may need to apply some geometric principle. Then read **_the question through a second time_** to pick out specific facts. Identify what is *given* and what you need to *find*.
2. **_Decide how you will solve the problem_**. You may need to use one of the special problem-solving strategies discussed in this section. If you decide to solve a word problem by using an algebraic method, you may need to first translate the conditions of the problem into an equation or an inequality.
3. **_Carry out your plan._**
4. **_Verify that your answer is correct_** by making sure it works in the original question.

A Brief Review of Key Math A Facts and Skills

1. NUMBER CONCEPTS

The Set of Real Numbers and Its Subsets

1.1 INTEGERS

- **Integers** include the counting numbers (positive integers), their opposites (negative integers), and 0:

$$\ldots, -4, -3, -2, -1, 0, 1, 2, 3, 4, \ldots$$

- When two integers are multiplied together, the answer is called the **product** and the numbers being multiplied together are **factors**. Since $2 \times 3 = 6$, 2 and 3 are factors of the product, 6.

- If $a \div b$ has a 0 remainder, then a is evenly divisible by b. Thus, 6 is divisible by 3 since $6 \div 3 = 2$ with remainder 0, but 6 is not divisible by 4 since $6 \div 4 = 1$ with remainder 2.

- **Even integers** are integers that are divisible by 2:

$$\ldots, -6, -4, -2, 0, 2, 4, 6, \ldots$$

- **Odd Integers** are integers that are *not* divisible by 2:

$$\ldots, -5, -3, -1, 1, 3, 5, \ldots$$

- **Prime numbers** are integers greater than 1 that are divisible only by themselves and 1:

$$2, 3, 5, 7, 11, 13, 17, \ldots$$

1.2 GENERAL INTEGERS

• If n represents an integer, then four consecutive integers are

$$n, n + 1, n + 2, n + 3$$

• If n represents an *even* integer, then four consecutive even integers are

$$n, n + 2, n + 4, n + 6$$

• If n represents an *odd* integer, then four consecutive odd integers are

$$n, n + 2, n + 4, n + 6$$

1.3 SQUARE ROOT RADICALS

The square root notation $\sqrt{n}$ means one of two equal nonnegative numbers whose product is n. For example, $\sqrt{16} = 4$ since $4 \times 4 = 16$. The number underneath the radical sign $\left(\sqrt{}\right)$ is called the **radicand**. Whole numbers such as 1, 4, 9, 16, 25, . . . are called **perfect squares** since their square roots are whole numbers. Fractions like $\dfrac{4}{9}$ with numerators and denominators that are both perfect squares are also perfect squares since their square roots are fractions with whole numbers in the numerators and denominators. For example, $\sqrt{\dfrac{4}{9}} = \dfrac{2}{3}$.

1.4 REAL NUMBERS

• A **rational number** is a number that can be written as a fraction with an integer numerator and a nonzero integer denominator. The set of rational numbers includes integers, decimal numbers such as 1.25 $\left(= \dfrac{5}{4}\right)$, and nonending decimal numbers in which one or more nonzero digits repeat endlessly, as in 0.3333. . . $\left(= \dfrac{1}{3}\right)$ and 0.636363. . . $\left(= \dfrac{7}{11}\right)$.

- The square root of a number that is *not* a perfect square, such as $\sqrt{5}$, is an **irrational number**. The quantity π ($= 3.1415926\ldots$) is another example of an irrational number. An irrational number does *not* have an *exact* decimal equivalent, but can be approximated using a calculator.
- The set of **real numbers** is the union of the set of rational numbers and the set of irrational numbers.

1.5 PROPERTIES OF REAL NUMBERS

- COMMUTATIVE PROPERTIES: The order in which two real numbers are added or multiplied does not matter. Thus:

$$x + y = y + x \quad \text{and} \quad xy = yx$$

- ASSOCIATIVE PROPERTIES: The order in which three real numbers are grouped when added or multiplied does not matter. Thus:

$$x + (y + z) = (x + y) + z \quad \text{and} \quad x(yz) = (xy)z$$

- IDENTITY PROPERTIES: For addition, 0 is the identity element since, for any real number x:

$$x + 0 = 0 + x$$

For multiplication, 1 is the identity element since, for any real number x:

$$x \cdot 1 = 1 \cdot x$$

- INVERSE PROPERTIES: For addition, the inverse of a real number x is its opposite, $-x$, since

$$x + (-x) = 0$$

For multiplication, the inverse of a nonzero real number x is its reciprocal, $\frac{1}{x}$, since

$$x \cdot \left(\frac{1}{x}\right) = 1$$

- DISTRIBUTIVE PROPERTY: For three real numbers x, y, and z:

$$x(y+z) = xy + yz$$

1.6 COMBINING LIKE TERMS

Like terms have the same variable factors but may have different numerical coefficients, as in $2xy$ and $3xy$. To combine like terms, use the reverse of the distributive property. For example, $5x + 4x = (5+4)x = 9x$. Thus, for like terms, combine the numerical coefficients and keep the common variable factors, as in $7ab - 3ab = 4ab$.

1.7 OPERATIONS WITH RADICALS

- To *simplify* a radical, rewrite the radicand as the product of two whole numbers one of which is the greatest possible perfect square factor of the product. Then write the radical over each factor and simplify. For example:

$$\sqrt{75} = \sqrt{25 \cdot 3} = \sqrt{25} \cdot \sqrt{3} = 5\sqrt{3}$$

- To *combine* radicals, first rewrite the radicals so that they have the same radicand, if possible. Then treat the radicals as like terms and combine. For example:

$$\sqrt{18} + 4\sqrt{2} = \sqrt{9 \cdot 2} + 4\sqrt{2} = 3\sqrt{2} + 4\sqrt{2} = 7\sqrt{2}$$

- To *multiply* or *divide* radicals, use these rules:

$$a\sqrt{b} \times c\sqrt{d} = a \cdot c\sqrt{b \cdot d} \quad \text{and} \quad \frac{a\sqrt{b}}{c\sqrt{d}} = \frac{a}{c}\sqrt{\frac{b}{d}}$$

1.8 PERCENT

The term *percent* means "parts of 100." For example, 60% means $\frac{60}{100}$ or 0.60 or $\frac{3}{5}$.

- To find the percent of a number, multiply the number by the percent in decimal form. For example, if the rate of sales tax is 6%, the amount of sales tax on a $40 item is 6% of $40 or $0.06 \times$ $40 = 2.40.

- To find what percent one number is of another, write the "is" number over the "of" number and multiply by 100%. For instance, if the amount of sales tax on a $65 item is $5.20, then the tax rate is obtained by finding what percent 5.20 is of 65. Since

$$\frac{5.20\,(\text{"is" number})}{65\,(\text{"of" number})} \times 100\% = \frac{520}{65}\,\%$$
$$= 8\%$$

The tax rate is 8%.

- To find a number when a percent of it is given, divide the given amount by the percent in decimal form. For example, if 30% of some number n is 12, then $n = \dfrac{12}{0.30} = 40$.

SIGNED NUMBERS

1.9 OPERATIONS WITH SIGNED NUMBERS

Real numbers are ordered so that each number on a number line is greater than the number to its left, as shown in the accompanying graph. Thus, $2 > 1$ and $-1 > -2$.

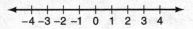

The **absolute value** of a number is its distance from 0 on the number line. Since distance cannot be a negative number, the absolute value of a number is always nonnegative. The notation $|n|$ represents the absolute value of n and is equal to the number n without its sign. Thus, $|-3| = 3$.

- To *multiply or divide* signed numbers with the *same* sign, perform the operation and attach a plus sign. For example:

$$(-3) \times (-5) = +15$$

- To *multiply or divide* signed numbers with *different* signs, perform the operation and attach a minus sign. For example:

$$\frac{+15}{-3} = -5$$

- To *add* signed numbers with the *same* sign, add the numbers and attach the common sign to the sum. For example:

$$(-3) + (-4) = -7$$

- To *add* signed numbers with *different* signs, subtract the unsigned numbers and attach to the sum the sign of the number with the larger absolute value. For example:

$$(-5) + (+2) = -(5 - 2) = -3$$

- To *subtract* signed numbers, change to an equivalent addition example by taking the opposite of the number that is being subtracted. For example:

$$(-7) - (-2) = (-7) + (+2) = -5$$

OPERATIONS WITH EXPONENTS

1.10 LAWS OF EXPONENTS

An **exponent** indicates how many times a number is to be used as a factor in a product. In 2^3, the exponent, 3, means that 2, called the *base*, should be used as a factor three times, as in $2 \times 2 \times 2 = 8$.

- Any quantity with a zero exponent is equal to 1. Thus, $5^0 = 1$.

- Any quantity with a negative exponent can be written with a positive exponent by inverting the base, as in

$$x^{-n} = \frac{1}{x^n} \quad (x \neq 0)$$

For example:

$$2^{-4} = \frac{1}{2^4} \quad \text{and} \quad \frac{1}{5^{-2}} = 5^2$$

- PRODUCT LAW: To **multiply** powers of the *same* base, *add* their exponents, as in

$$y^a \times y^b = y^{a+b}$$

For example:

$$2^3 \times 2^4 = 2^{3+4} = 2^7 \quad \text{and} \quad 3x^2 \cdot 2x^5 = (3 \cdot 2)(x^2 \cdot x^5) = 6x^7$$

- QUOTIENT LAW: To **divide** powers of the *same* base, *subtract* their exponents, as in

$$\frac{y^a}{y^b} = y^{a-b}$$

For example:

$$\frac{2^7}{2^3} = 2^{7-3} = 2^4 \quad \text{and} \quad \frac{8xy^3}{2y} = \frac{8x}{2} \cdot \frac{y^3}{y^1} = 4xy^2$$

- POWER LAW: To raise a power to another power, multiply the exponents, as in

$$(y^a)^b = y^{a \times b}$$

For example:

$$(2^3)^4 = 2^{3 \times 4} = 2^{12}$$

1.11 SCIENTIFIC NOTATION

A number is in **scientific notation** when it is written as the product of a number between 1 and 10 and a power of 10.

- When a number greater than 10 is written in scientific notation, the power of 10 will be the number of places the decimal point must be moved to the *left*. For example:

$$4\,7\,0,0\,0\,0. = 4.7 \times 10^5$$

+5

- When a number between 0 and 1 is written in scientific notation, the power of 10 will be the number of places the decimal point must be moved to the *right*, with a negative sign in front of the exponent. For example:

$$0.0\,0\,0\,0\,0\,3\,0\,9 = 3.09 \times 10^{-6}$$

−6

Practice Exercises

1. Which property is illustrated by the following equation?

$$\square(\triangle + O) = \square\triangle + \square O$$

(1) distributive   (3) commutative
(2) associative   (4) transitive

2. Which equation illustrates the additive inverse property?

(1) $a + (-a) = 0$ (3) $a \div (-a) = -1$

(2) $a + 0 = a$ (4) $a \cdot \dfrac{1}{a} = 1$

3. In the solution of the equation $3x = 6$, which property of real numbers justifies statement 5?

Statements	Reasons
1. $3x = 6$	1. Given
2. $\frac{1}{3}(3x) = \frac{1}{3}(6)$	2. Multiplication axiom
3. $(\frac{1}{3} \cdot 3)x = 2$	3. Associative property
4. $1 \cdot x = 2$	4. Multiplicative inverse
5. $x = 2$	5. _____

(1) Closure   (3) Commutative
(2) Identity   (4) Inverse

4. The expression $\sqrt{300}$ is equivalent to

(1) $50\sqrt{6}$ (3) $3\sqrt{10}$

(2) $12\sqrt{5}$ (4) $10\sqrt{3}$

5. The expression $\sqrt{27} + \sqrt{12}$ is equivalent to

(1) $\sqrt{39}$ (3) $5\sqrt{6}$

(2) $13\sqrt{3}$ (4) $5\sqrt{3}$

6. The sum of $2\sqrt{3}$ and $\sqrt{12}$ is

(1) $4\sqrt{6}$ (3) $3\sqrt{15}$

(2) $8\sqrt{3}$ (4) $4\sqrt{3}$

7. The expression $\sqrt{50} + 3\sqrt{2}$ can be written in the form $x\sqrt{2}$. Find x.

8. If 0.00037 is expressed as 3.7×10^n, what is the value of n?

9. When expressed in scientific notation, the number $0.0000000364 = 3.64 \times 10^n$. The value of n is

(1) 8   (3) -10

(2) 10   (4) -8

10. The number 0.00000467 can be written in the form 4.67×10^n. Find n.

Solutions

1. If a, b, and c, are real numbers, the distributive property states that

$$a(b + c) = ab + bc$$

The given equation

$$\square(\triangle + O) = \square\triangle + \square O$$

has the form $a(b + c) = ab + bc$, where $a = \square$, $b = \triangle$, and $c = O$.

The correct choice is **(1)**.

2. The equation $a + (-a) = 0$ states that the sum of any number a and its additive inverse (opposite) is 0.

The correct choice is **(1)**.

3. The number 1 is the *identity element* for multiplication for the set of real numbers since the product of 1 and *any* real number is the same real number.

Statement 4 is $1 \cdot x = 2$. Since 1 is the identity element for multiplication, the left side of the equation in statement 4 becomes x. Then statement 5, $x = 2$, follows from statement 4 as a result of 1 being the identity element for multiplication.

The correct choice is **(2)**.

4. The given expression is: $\sqrt{300}$

Simplify $\sqrt{300}$ by finding two factors of the radicand, 300, one of which is the highest perfect square that divides into 300: $\sqrt{100(3)}$

Take the square root of the perfect square factor, 100, and place it outside the radical sign as the coefficient, 10: $10\sqrt{3}$

The correct choice is **(4)**.

5. The given expression is:

$$\sqrt{27} + \sqrt{12}$$

You cannot add $\sqrt{27}$ and $\sqrt{12}$ in their present form because only *like radicals* may be combined. Like radicals have the same radicand (number under the radical sign) and the same index (here, understood to be 2, representing the square root). A radical can be simplified by finding two factors of the radicand, one of which is the highest possible perfect square.

Factor the radicands, using the highest possible perfect square in each:

$$\sqrt{9 \cdot 3} + \sqrt{4 \cdot 3}$$

Simplify by taking the square root of the perfect square factor and placing it outside the radical sign as a coefficient of the radical:

$$3\sqrt{3} + 2\sqrt{3}$$

Combine the like radicals by adding their coefficients to get the new coefficient:

$$5\sqrt{3}$$

The given expression is equivalent to $5\sqrt{3}$.

The correct choice is **(4)**.

6. The given expression is:

$$2\sqrt{3} + \sqrt{12}$$

Only *like radicals* can be added. Like radicals must have the same root (here both are square roots) and must have the same radicand (the number under the radical sign). Factor out any perfect square factor in the radicands:

$$2\sqrt{3} + \sqrt{4(3)}$$

Remove the perfect square factors from under the radical sign by taking their square roots and writing them as coefficients of the radical:

$$2\sqrt{3} + 2\sqrt{3}$$

Combine the like radicals by adding their coefficients and writing the sum as the numerical coefficient of the common radical:

$$4\sqrt{3}$$

The correct choice is **(4)**.

7. The given expression is: $\sqrt{50} + 3\sqrt{2}$

Factor out any perfect square factors in the radicand (the number under the radical sign): $\sqrt{25(2)} + 3\sqrt{2}$

Remove the perfect square factors from under the radical sign by taking their square roots and writing them as coefficients of the radical: $5\sqrt{2} + 3\sqrt{2}$

Combine the like radicals by combining their coefficients and using the sum as the numerical coefficient of the common radical: $8\sqrt{2}$

$8\sqrt{2}$ is in the form $x\sqrt{2}$ with $x = 8$.

$x = \mathbf{8}$.

8. To change 3.7 to 0.00037, the decimal point in 3.7 must be moved four places to the left. Each move of one place to the left is equivalent to dividing by 10. Therefore, 3.7 must be divided by 10^4 to equal 0.00037. Division by 10^4 is equivalent to multiplying by $\dfrac{1}{10^4}$ or by 10^{-4}. Therefore, $0.00037 = 3.7 \times 10^{-4}$, so $n = -4$.

$n = \mathbf{-4}$.

9. The given equation is $0.0000000364 = 3.64 \times 10^n$.

In order for 3.64×10^n to become 0.0000000364, the decimal point in 3.64 must be moved eight places to the left. Each move of one place to the left is equivalent to dividing 3.64 by 10, or equivalent to multiplying it by 10^{-1}. Therefore, to move the decimal point eight places to the left, 3.64 must be multiplied by 10^{-8}, that is, $n = -8$.

The correct choice is (**4**).

10. Dividing 4.67 by 10 moves the decimal point one place to the left. Therefore, 4.67 must be divided by 10 six times to make it equal 0.00000467. Dividing by 10 six times is the same as multiplying by 10^{-6}: $0.00000467 = 4.67 \times 10^{-6}$.

$n = \mathbf{-6}$.

2. LOGIC

2.1 OPEN SENTENCES VERSUS STATEMENTS

The equation $2x + 1 = 9$ is an **open sentence** since it cannot be judged true or false until the variable x is replaced by a number. The number replacement that makes the equation $2x + 1 = 9$ a true statement is called a **root** or **solution** of the equation. In logic, a **statement** is a sentence that can be judged to be either true or false.

- In algebra, a single letter such as x is used to represent an unknown number. In logic, a single letter such as p is used to represent an unknown statement.
- A statement p and its *negation*, written as $\sim p$, have opposite truth values. For example:

p: I am hungry.
negation of p: I am *not* hungry.

If statement p (I am hungry) is true, then the negation of statement $\sim p$ (I am *not* hungry) is false.

2.2 COMPOUND STATEMENTS

Two simple statements can be joined together to form a **compound statement** by using *logical connectives*. The accompanying table summarizes the four logical connectives, where p and q represent simple statements.

Compound Statement	Form	Truth Value
Conjunction	p AND q	True when p and q are both true.
Disjunction	p OR q	True when p is true, q is true, or both p and q are true.
Conditional	IF p, THEN q.	True *except* when p is true and q is false.
Biconditional	p IF and only IF q.	True when p and q are both true or when both are false.

2.3 RELATED CONDITIONALS

A **conditional** statement has the form "If <u>statement 1</u>, then <u>statement 2</u>." From this conditional statement, three related conditional statements can be formed: the *converse, inverse*, and *contrapositive*.

- To form the **converse** of a conditional statement, switch the first and second statements: "If <u>statement 2</u>, then <u>statement 1</u>."
- To form the **inverse** of a conditional statement, negate both parts of the original conditional: "If <u>*not* statement 1</u>, then <u>*not* statement 2</u>."
- To form the **contrapositive** of a conditional statement, negate and then switch both parts of the original conditional: "If <u>*not* statement 2</u>, then <u>*not* statement 1</u>."

2.4 LOGICALLY EQUIVALENT STATEMENTS

Two statements are logically equivalent if they always have the same truth value. A conditional statement and its contrapositive are **logically equivalent** statements. A conditional statement and its converse do *not* necessarily have the same truth value, so they are *not* logically equivalent statements.

2.5 DRAWING INFERENCES

Suppose p and q represent simple statements so that the disjunction "p OR q" is a true statement.

- If it is also known that p is false, then it must be the case that q is true.
- Similarly, if it is known that q is false, then it must be the case that p is true.

Now suppose p and q represent simple statements so that the conditional statement "If p, then q" is a true statement.

- If it is also known that p is true, then you can infer that q must also be true.
- If it is also known that q is false, then you can infer that p must also be false.

Practice Exercises

1. Which statement is logically equivalent to the statement "If we recycle, then the amount of trash in landfills is reduced"?

   (1) If we do not recycle, then the amount of trash in landfills is not reduced.
   (2) If the amount of trash in landfills is not reduced, then we do not recycle.
   (3) If the amount of trash in landfills is reduced, then we recycle.
   (4) If we do not recycle, then the amount of trash in landfills is reduced.

2. When the statement "If A, then B" is true, which statement *must* also be true?

   (1) If B, then A.
   (2) If not A, then B.
   (3) If not B, then A.
   (4) If not B, then not A.

Solutions

1. Two statements are logically equivalent if they always have the same truth value.

A conditional, $p \rightarrow q$, and its contrapositive, $\sim q \rightarrow \sim p$, are logically equivalent statements. To form the contrapositive of the given conditional statement, "If we recycle, then the amount of trash in landfills is reduced," interchange the hypothesis and the conclusion and then negate both parts of the conditional.

The contrapositive of the given condition is "If the amount of trash in landfills is not reduced, then we do not recycle."

The correct choice is (**2**).

2. A conditional statement and its contrapositive always have the same truth value.

To form the contrapositive of a conditional statement, interchange the two parts of the conditional statement and then negate both parts. For the original statement "If A, then B," the contrapositive is "If not B, then not A."

Thus, when the given statement, "If A, then B," is true, the statement "If not B, then not A" must also be true.

The correct choice is (**4**).

3. ALGEBRAIC METHODS

Operations with Polynomials

3.1 MONOMIALS

A **monomial** is a single term containing a number, a variable, or the product of numbers and variables.

- To *multiply* monomials, multiply the numerical coefficients and multiply like variable factors by adding their exponents. For example:

$$(-4a^2b)(2a^3b) = -8a^{2+3}b^{1+1} = -8a^5b^2$$

- To *divide* monomials, divide the numerical coefficients and divide like variable factors by subtracting their exponents. For example:

$$\frac{-8a^5b^2}{2a^3b} = \left(\frac{-8}{2}\right)\left(\frac{a^5}{a^3}\right)\left(\frac{b^2}{b}\right)$$
$$= -4a^{5-3}b^{2-1}$$
$$= -4a^2b$$

3.2 POLYNOMIALS

A **polynomial** is the sum of two or more unlike monomials.

- To *add* two polynomials, write one polynomial underneath the other with like terms aligned in the same column. Then combine like terms. For example:

$$3x^2 + x + 8$$
$$+ x^2 \qquad - 9$$
$$\overline{4x^2 + x - 1}$$

- To *subtract* one polynomial from another polynomial, change to an addition example by taking the opposite of each term of the polynomial that is being subtracted. For example, to subtract $x^2 - 9$ from $4x^2 + x - 1$, write the opposite of each term of $x^2 - 9$ under the corresponding term in $4x^2 + x - 1$. Then add like terms.

$$4x^2 + x - 1$$
$$-x^2 \quad\quad + 9$$
$$\overline{}$$
$$3x^2 + x + 8$$

- To *multiply* a polynomial by a monomial, use the distributive property. For example:

$$3x(-2x^3 + 5x^2 - 8) = 3x(-2x^3) + 3x(5x^2) + 3x(-8x)$$
$$= -6x^4 \quad\quad + 15x^3 \quad -24x^2$$

- To multiply two binomials together, use "FOIL." For example:

$$\overset{\textit{First terms}}{\overbrace{}} \quad \overset{\textit{Outer terms}}{\overbrace{}} \quad \overset{\textit{Inner terms}}{\overbrace{}} \quad \overset{\textit{Last terms}}{\overbrace{}}$$
$$(x + 5)(x - 3) = \overbrace{(x)(x)} + \overbrace{(-3)(x)} + \overbrace{(5)(x)} + \overbrace{(5)(-3)}$$
$$= x^2 + 2x - 15$$

- To *divide* a polynomial by a monomial, divide each term of the polynomial by the monomial. For example:

$$\frac{12z^4 + 20z^3 - 4z^2}{-4z^2} = \frac{12z^4}{-4z^2} + \frac{20z^3}{-4z^2} + \frac{-4z^2}{-4z^2}$$
$$= -3z^2 - 5z \quad + 1$$

3.3 FACTORING

Factoring undoes multiplication.

- Factoring out the greatest common factor of the terms of a polynomial reverses the distributive property. For example:

$$x^2 + 3x = x(x + 3)$$

and

$$2y^3 - 8y^2 + 6y = 2y(y^2 - 4y + 3)$$

You can check that you have factored correctly by multiplying the two factors together and verifying that the product is the original polynomial.

- The difference of the squares of two quantities can be factored as the product of the sum and difference of the terms that are being squared. For example:

$$x^2 - 4 = (x)^2 - (2)^2 = (x + 2)(x - 2)$$

and

$$2y^3 - 18y = 2y(y^2 - 9) = 2y(y + 3)(y - 3)$$

- Quadratic trinomials that may appear on the Math A Regents exam can be factored using the reverse of FOIL. For example:

$$x^2 + 2x - 15 = (x + a)(x + b)$$

where a and b are chosen so that $a \times b = -15$ and, at the same time, $a + b = +2$. Since $(+5) \times (-3) = -15$ and $(+5) + (-3) = +2$:

$$x^2 + 2x - 15 = (x + 5)(x - 3)$$

Operations with Algebraic Fractions

3.4 ALGEBRAIC FRACTIONS

If a variable is in the denominator of a fraction, you can assume that it cannot represent a number that would make the denominator evaluate to 0. For example, in the fraction $\dfrac{x}{x - 2}$, x cannot be equal to 2 since $2 - 2 = 0$ and a fraction with a zero denominator is not defined.

- To *simplify* a fraction, factor the numerator and factor the denominator. Then divide out any factor that is contained in the numerator and in the denominator since their quotient is 1. For example:

$$\frac{2x^2y - 50y}{4x^2 + 20x} = \frac{2y(x^2 - 25)}{4x(x + 5)}$$

$$= \frac{\overset{1}{2y(x + 5)}(x - 5)}{\underset{2}{4x(x + 5)}}$$

$$= \frac{y(x - 5)}{2x}$$

- To *multiply* algebraic fractions, first divide out any factor that is common to a numerator and a denominator. Then write the product of the remaining factors of the numerators over the product of the remaining factors of the denominators. For example:

$$\frac{2x^3}{x^2-x-12} \cdot \frac{x^2-16}{6x} = \frac{x^2 \cdot 2x}{(x+3)(x-4)} \cdot \frac{(x-4)(x+4)}{6x}$$

$$= \frac{x^2 \cdot 2x}{(x+3)(x-4)} \cdot \frac{\overset{1}{\cancel{(x-4)}}(x+4)}{\underset{3}{\cancel{6x}}}$$

$$= \frac{x^2(x+4)}{3(x+3)}$$

- To *divide* algebraic fractions, change to a multiplication example by inverting the second fraction.
- To *combine* fractions with *like* denominators, write the sum of their numerators over their common denominator. For example:

$$\frac{5x+1}{3} + \frac{x-13}{3} = \frac{(5x-1)+(x+13)}{3}$$

$$= \frac{6x-12}{3}$$

$$= \frac{\overset{2}{\cancel{6}}(x-2)}{\cancel{3}}$$

$$= 2(x-2)$$

- To *combine* fractions with *unlike* denominators, first change each fraction into an equivalent fraction with the least common denominator (LCD) of the fractions as its denominator. Then follow the rules for combining fractions with like denominators. For example, the LCD of the fractions in the sum $\frac{x+2}{2x} + \frac{1}{3x}$ is $6x$ since that is the smallest expression into which both denominators divide evenly. Since $6x \div 2x = 3$ and $6x \div 3x = 2$, multiply the first fraction by 1 in the form of $\frac{3}{3}$ and multiply the second fraction by 1 in the

form of $\frac{2}{2}$:

$$\frac{x+2}{2x} + \frac{1}{3x} = \frac{3}{3}\left(\frac{x+2}{2x}\right) + \frac{2}{2}\left(\frac{1}{3x}\right)$$
$$= \frac{3(x+2) + 2(1)}{6x}$$
$$= \frac{3x + 6 + 3}{6x}$$
$$= \frac{3x + 8}{6x}$$

Solving Equations

3.5 LINEAR EQUATIONS

An equation such as $2x - 3 = 7x$ is a **first-degree** (or **linear**) **equation** since the greatest exponent of the variable is 1.

- To solve a first-degree equation in one variable, isolate the variable by undoing operations that are connected to the variable. When undoing an operation on one side of the equation, do the same thing on the opposite side. For example, if $3x = 2.1$, undo the multiplication of x by 3 by *dividing* both sides of the equation by 3: $\frac{3x}{3} = \frac{2.1}{3}$, so $x = 0.7$. If an equation contains parentheses, remove them first. For example, to solve $3(x + 1) - x = 13$ for x:

Remove the parentheses: $\qquad\qquad\qquad\qquad 3x + 3 - x = 13$
Collect like terms on the same side of
the equation: $\qquad\qquad\qquad\qquad\qquad\qquad 2x = 13 - 3$

Divide both sides of the equation by 2: $\qquad\qquad x = \frac{10}{2} = 5$

- To solve an equation for a given variable in terms of another variable, isolate the given letter. For example, to solve $x + y = 7y - x$ for x, rearrange the terms of the equation so that like letters appear on the same side of the equation:

Add x to each side: $\qquad\qquad\qquad\qquad 2x + y = 7y$

Subtract y from each side: $\qquad\qquad\qquad\quad 2x = 6y$

Divide each side by 2: $\qquad\qquad\qquad\qquad\quad x = 3y$

3.6 RATIO AND PROPORTION

- A ratio compares two quantities by division. The ratio of x to y is written as $x{:}y$ or as $\dfrac{x}{y}$. If x is three times as great as y, then

$\dfrac{x}{y} = \dfrac{3}{1}$ or 3:1.

- A proportion is an equation that states that two ratios are numerically equal. To change the proportion $\dfrac{a}{b} = \dfrac{c}{d}$ into an equation without fractions, cross-multiply: $\dfrac{a}{b} = \dfrac{c}{d}$ becomes $b \times c = a \times d$.

For example, if $\dfrac{x}{x+4} = \dfrac{3}{5}$, then after cross-multiplying:

$$5x = 3(x + 4)$$
$$5x = 3x + 12$$
$$2x = 12$$
$$x = \frac{12}{2} = 6$$

3.7 LINEAR INEQUALITIES

The symbols for inequality comparisons are as follows:

$<$ means "is less than," as in $2 < 5$.

$>$ means "is greater than," as in $7 > 3$.

$\leq$ means "is less than *or* equal to."

$\geq$ means "is greater than *or* equal to."

- A first-degree inequality is solved in much the same way that a first-degree equation is solved except that, when *multiplying* or *dividing* both sides of an inequality by the same *negative* number, you must *reverse* the inequality. For example, to solve $1 - 2x > 9$ for x, first subtract 1 from each side to obtain $-2x > 8$. Then divide each side by -2 and, at the same time, reverse the inequality:

$$\frac{-2x}{-2} < \frac{8}{-2}$$

$$x < -4$$

- Solve a combined inequality such as $-3 \leq 2x - 1 < 9$ by isolating the variable in the middle part of the inequality:

Add 1 to each member: $\qquad\qquad -2 \leq 2x < 10$

Divide each member by 2: $\qquad \dfrac{-2}{2} \leq \dfrac{2x}{2} < \dfrac{10}{2}$

Simplify: $\qquad\qquad\qquad\qquad -1 \leq x < 5$

The inequality $-1 \leq x < 5$ includes all numbers from -1 (including -1) to 5 (not including 5), as shown in the accompanying number line. A darkened circle around an endpoint indicates that the point is included in the interval, while an unshaded circle around an endpoint means that the point is *not* included.

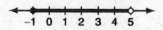

3.8 FRACTIONAL EQUATIONS

To solve an equation with fractions, clear the equation of its fractions by multiplying each member by the least common denominator of the fractional terms. For example:

The given equation is: $\qquad\qquad \dfrac{2x}{x} + \dfrac{1}{7} = \dfrac{4}{x}$

Multiply each term of the equation by $7x$, the lowest common multiple of the denominators:

$$7x\left(\frac{2}{x}\right) + 7x\left(\frac{1}{7}\right) = 7x\left(\frac{4}{x}\right)$$

Subtract 14 from each side of the equation:

$$\begin{array}{rcr} 14 + x & = & 28 \\ -14 & = & -14 \\ \hline x = & & 14 \end{array}$$

The value of x is **14**.

3.9 QUADRATIC EQUATIONS

To solve a factorable quadratic equation in which 0 appears alone on one side of the equation, factor the other side. Set each factor equal to 0, and solve the two equations that result.

- If $x^2 + 3x = 0$, then $x(x + 3) = 0$, so $x = 0$ or $x + 3 = 0$, making $x = 0$ or $x = -3$.
- If $x^2 - 4x + 4 = 0$, then $(x - 2)(x - 2) = 0$, so $x - 2 = 0$, making $x = 2$.
- If $x^2 + 2x = 15$, subtract 15 from each side of the equation so that all of the nonzero terms are on the left side, as in $x^2 + 2x - 15 = 0$. Then factor: $(x + 5)(x - 3) = 0$, so $x + 5 = 0$ or $x - 3 = 0$, making $x = -5$ or $x = 3$.

Practice Exercises

1. Express $\dfrac{15x^2}{-3x}$ in simplest form.

2. The expression $(3x^2y^3)^2$ is equivalent to
   (1) $9x^4y^6$ (2) $9x^4y^5$ (3) $3x^4y^6$ (4) $6x^4y^6$

3. The product of $(-2xy^2)(3x^2y^3)$ is
   (1) $-5x^3y^5$ (2) $-6x^2y^6$ (3) $-6x^3y^5$ (4) $-6x^3y^6$

4. Subtract $4m - h$ from $4m + h$.

5. Find the sum of $2x^2 + 3x - 1$ and $3x^2 - 2x + 4$.

6. Express the product $(2x - 7)(x + 3)$ as a trinomial.

7. When $3x^3 + 3x$ is divided by $3x$, the quotient is
   (1) x^2 (2) $x^2 + 1$ (3) $x^2 + 3x$ (4) $3x^3$

8. Solve for y: $6(y + 3) = 2y - 2$.

9. Solve for y: $\dfrac{y}{3} + 2 = 5$.

10. Factor: $x^2 - 49$.

11. Factor: $x^2 + x - 30$.

12. Factor: $x^2 + 5x - 14$.

13. The solution set of $x^2 - x - 6 = 0$ is
   (1) {1,–6}   (2) {–3,2}   (3) {3,–2}   (4) {5,1}

14. In factored form, the trinomial $3x^2 + 5x - 2$ is equivalent to

   (1) $(3x + 1)(x - 2)$ (3) $(3x + 2)(x - 1)$
   (2) $(3x - 1)(x + 2)$ (3) $(3x - 2)(x + 1)$

15. Express the quotient in simplest form:

$$\frac{x^2 - 4}{x^2 + 3x - 10} \div \frac{x^2 + 5x + 6}{x^2 + 8x + 15}$$

16. *a.* Express in simplest form:

$$\frac{5}{x-1} - \frac{3}{x} \quad (x \neq 1, 0)$$

   b. The numerator of a certain fraction is 3 less than the denominator. If the numerator and the denominator are each increased by 1, the value of the fractions is $\frac{2}{3}$. Find the original fraction.

Solutions

1. The given expression is:

$$\frac{15x^2}{-3x}$$

To divide two monomials, first divide their numerical coefficients to find the numerical coefficient of the quotient:

$$(15) \div (-3) = -5$$

Divide the literal factors to find the literal factor of the quotient. Remember that powers of the same base are divided by subtracting their exponents:

$$(x^2) \div (x^1) = x^1$$

Combine the two results:

$$\frac{15x^2}{-3x} = -5x$$

The fraction in simplest form is **$-5x$**.

2. The given expression, $(3x^2y^3)^2$, means $(3x^2y^3)(3x^2y^3)$.

The numerical coefficient of the product is the product of the two numerical coefficients:

$$(3)(3) = 9$$

In multiplying powers of the same base, add the exponents to obtain the exponent for that base in the product:

$$(x^2y^3)(x^2y^3) = x^4y^6$$

Hence:

$$(3x^2y^3)^2 = 9x^4y^6$$

The correct choice is (**1**).

3. The given expression is:

$$(-2xy^2)(3x^2y^3)$$

To multiply two monomials, first find the numerical coefficient of the product by multiplying the two numerical coefficients together:

$$(-2)(3) = -6$$

Find the literal factor of the product by multiplying the literal factors together. The product of two powers of the same base is found by adding the exponents of that base:

$$(x^1y^2)(x^2y^3) = x^3y^5$$

Combine the two results:

$$(-2xy^2)(3x^2y^3) = -6x^3y^5$$

The correct choice is (**3**).

4. Subtraction is the inverse operation of addition. Thus, to subtract, add the additive inverse of the subtrahend (expression to be subtracted) to the minuend (expression subtracted from). Therefore, to subtract $4m - h$ from $4m + h$, add $-4m + h$ to $4m + h$:

$$\text{Subtract:} \quad \begin{array}{r} 4m + h \\ \underline{4m - h} \end{array} \qquad \text{means add:} \quad \begin{array}{r} 4m + h \\ \underline{-4m + h} \\ 2h \end{array}$$

The difference is **$2h$**.

5. Write the second trinomial under the first, placing like terms in the same column:

$$\begin{array}{r} 2x^2 + 3x^2 - 1 \\ \underline{3x^2 - 3x^2 + 1} \end{array}$$

Add each column by adding the numerical coefficients algebraically and bringing down the literal factor:

$$5x^2 + x + 3$$

The sum is **$5x^2 + x^2 + 3$**.

6. The product of two binomials may be found by multiplying each term of one by each term of the other, using a procedure analogous to that used in arithmetic to multiply multidigit numbers:

$$\begin{array}{r} 2x - 7 \\ \underline{x + 3} \\ 2x^2 - 7x \\ \underline{6x - 21} \\ 2x^2 - x - 21 \end{array}$$

Combine like terms:

The trinomial is **$2x^2 - x - 21$**.

7. Indicate the division in fractional form: Apply the distributive law by dividing each term of $3x^3 + 3x$ in turn by $3x$. Remember that, in dividing powers of the same base, the exponents are subtracted. Also note that $3x \div 3x = 1$:

$$\frac{3x^3 + 3x}{3x}$$

$$x^2 + 1$$

The correct choice is **(2)**.

8. The given expression is:
$$6(y + 3) = 2y - 2$$

Remove parentheses by applying the distributive law of multiplication over addition:
$$6y + 18 = 2y - 2$$

Add −18 (the additive inverse of 18) and also add −2y (the additive inverse of 2y) to both sides of the equation:
$$\frac{-2y - 18 = -2y - 18}{4y \qquad = \qquad -20}$$

Divide both sides by 4:
$$\frac{4y}{4} = \frac{-20}{4}$$

The solution is **y = −5**.
$$y = -5$$

9. The given expression is:
$$\frac{y}{3} + 2 = 5$$

To clear fractions in the equation, multiply each term by the least common denominator, in this case, by 3:
$$3\left(\frac{y}{3}\right) + 3(2) = 3(5)$$

Simplify:
$$y + 6 = 15$$

Add −6 (the additive inverse of 6) to both sides:
$$\frac{-6 = -6}{y \qquad = 9}$$

The solution is **y = 9**.

10. The given binomial, $x^2 - 49$ represents the difference between two perfect squares, x^2 and 49. To factor such an expression, take the square root of each perfect square:
$$\sqrt{x^2} = x \quad \text{and} \quad \sqrt{49} = 7$$

One factor will be the sum of the respective square roots, and the other factor will be the difference of the square roots:
$$x^2 - 49 = (x + 7)(x - 7)$$

The factored form is **(x + 7)(x − 7)**.

11. The given expression is a quadratic trinomial:

$$x^2 + x - 30$$

The factors of a quadratic trinomial are two binomials. The factors of the first term, x^2, are x and x, and they become the first terms of the binomials:

$$(x \quad)(x \quad)$$

The factors of the last term, -30, become the second terms of the binomials, but they must be chosen in such a way that the sum of the product of the inner terms and the product of the outer terms is equal to the middle term, $+x$, of the original trinomial. Try $+6$ and -5 as the factors of -30:

$$+6x = \text{inner product}$$
$$(x + 6)(x - 5)$$

$$-5x = \text{outer product}$$

Since $+6x$ and $-5x$ add up to $+x$, these are the correct factors:

$$(x + 6)(x - 5)$$

The factored form is $(x + 6)(x - 5)$.

12. The given expression is a quadratic trinomial:

$$x^2 + 5x - 14$$

The factors of a quadratic trinomial are two binomials. The first terms of the binomial are the factors, x and x, of the first term, x^2, of the trinomial:

$$(x \quad)(x \quad)$$

The second terms of the binomials are the factors of the last term, -14, of the trinomial. These factors must be chosen in such a way that the sum of the product of the inner terms and the product of the outer terms is equal to the middle term, $+5x$, of the original trinomial. The term

−14 has factors of 14 and −1, −14 and 1, +7 and −2, and −7 and +2. Try +7 and −2:

$+7x = $ inner product

$$(x + \overset{\frown}{7)(x} - 2)$$

$-2x = $ outer product

Since $(+7x) + (-2x) = +5x$, these are the correct factors:

$(x + 7)(x - 2)$

The factored form is $(x + 7)(x - 2)$.

13. The given expression is a quadratic equation:

$x^2 - x - 6 = 0$

The solution set is found by solving the equation by factoring. The left side is a *quadratic trinomial*, which can be factored into two binomials.

The factors of the first term, x^2, are x and x, and they constitute the first terms of each binomial factor:

$(x \quad)(x \quad) = 0$

The last term, −6, must be factored into two factors that will be the second terms of the binomials. The factors must be chosen in such a way that the sum of the product of the two inner terms and the product of the two outer terms is equal to the middle term of the original trinomial. Try $-6 = (-3)(+2)$:

$-3x = $ inner product

$$(x - \overset{\frown}{3)(x} + 2) = 0$$

$+2x = $ outer product

The sum of the inner and outer products is $-3x + 2x$ or $-x$, which is equal to the middle term of the original trinomial. Thus, the correct factors have been chosen, and the equation can be written as:

$(x - 3)(x + 2) = 0$

Since the product of two factors is 0, either factor may be equal to 0: $x - 3 = 0 \; or \; x + 2 = 0$

Add the appropriate additive inverse to each side, +3 in the case of the left equation and –2 in the case of the right one:

$$\begin{array}{cc} 3 = 3 & -2 = -2 \\ \hline x = 3 & x = -2 \end{array}$$

Thus, the solution set is {3,–2}.

The correct choice is **(3)**.

14. The given trinomial, $3x^2 + 5x - 2$, can be factored as the product of two binomials. Since the product of the first terms of the binomial factors must equal $3x^2$, the binomial factors must take the following form:

$$3x^2 + 5x - 2 = (3x + ?)(x + ?)$$

The missing terms in the two binomial factors must be a pair of numbers whose product is the last term of the trinomial, –2. Possible pairs of such numbers are –1, 2 and 1, –2.

The pair of numbers must be chosen and placed within the binomial factors in such a way that the sum of the inner and outer cross-products of the binomial factors equals the middle term, +5x, of the trinomial.

Try –1 and 2, and place them within the binomials as follows

$$-x = \text{inner product}$$

$$(3x - 1)(x + 2)$$

$$6x = \text{outer product}$$

Since $-x + 6x = +5x$, $(3x - 1)$ and $(x + 2)$ are the correct binomial factors of $3x^2 + 5x - 2$.

Hence, $3x^2 + 5x - 2 = (3x - 1)(x + 2)$.

The correct choice is **(2)**.

Alternative Solution: The problem can also be solved by using FOIL to express the product in each of the four choices as a trinomial. The correct choice is the one that gives a product that matches the given trinomial.

The correct choice is **(2)**.

15. The given expression is:

$$\frac{x^2-4}{x^2+3x-10} \div \frac{x^2+5x+6}{x^2+8x+15}$$

Change the division by a fractional expression to multiplication by inverting the divisor:

$$\frac{x^2-4}{x^2+3x-10} \cdot \frac{x^2+8x+15}{x^2+5x+6}$$

Factor each numerator and denominator wherever possible. The first numerator is a difference between two perfect squares. The other numerator and both denominators are quadratic trinomials that can be factored:

$$\frac{(x+2)(x-2)}{(x+5)(x-2)} \cdot \frac{(x+5)(x+3)}{(x+3)(x+2)}$$

If the same factor appears in both a numerator and a denominator, divide both of them by that factor (cancel):

$$\frac{\overset{1}{\cancel{(x+2)}}\,\overset{1}{\cancel{(x-2)}}}{\underset{1}{\cancel{(x+5)}}\,\underset{1}{\cancel{(x-2)}}} \cdot \frac{\overset{1}{\cancel{(x+5)}}\,\overset{1}{\cancel{(x+3)}}}{\underset{1}{\cancel{(x+3)}}\,\underset{1}{\cancel{(x+2)}}}$$

Multiply together the remaining factors in the numerator, and multiply together the remaining factors in the denominator:

$$\frac{1}{1}$$

The simplest form is:

$$1$$

The quotient in simplest form is **1**.

16. a. The given expression is:
In their present form, the fractions cannot be combined because they have different denominators. Express each fraction as an equivalent fraction

$$\frac{5}{x-1} - \frac{3}{x} \qquad (x \neq 1, 0)$$

having the least common denominator (LCD). The LCD is the simplest expression into which each of the denominators can be divided. The LCD for $(x - 1)$ and x is $x(x - 1)$. Multiply the first fraction by 1 in the form $\frac{x}{x}$, and multiply the second fraction by 1 in the form $\frac{x-1}{x-1}$:

$$\frac{5x}{x(x-1)} - \frac{3(x-1)}{x(x-1)}$$

NOTE: The above step is possible because $x \neq 1, 0$ ensures that you are not multiplying by 0:

$$\frac{5x}{x(x-1)} - \frac{3x-3}{x(x-1)}$$

Since the fractions now have the same denominator, they may be combined by combining their numerators:

$$\frac{5x-(3x-3)}{x(x-1)}$$

Remove the parentheses in the numerator:

$$\frac{5x-3x-3}{x(x-1)}$$

Combine like terms in the numerator:

$$\frac{2x+3}{x(x-1)}$$

The expression in simplest form is $\frac{2x+3}{x(x-1)}$.

b. Let x = denominator of the original fraction.

Then $x - 3$ = numerator.

If the numerator and denominator are each increased by 1, the value of the fraction is $\frac{2}{3}$:

$$\frac{x-2}{x+1} = \frac{2}{3}$$

The equation is in the form of a proportion. In a proportion, the product of

the means equals the product of the
extremes (cross-multiply):

$$3(x-2) = 2(x+1)$$
$$3x - 6 = 2x + 2$$
$$3x - 2x = 2 + 6$$

The denominator is represented by x: $x = 8$

The numerator is represented by $x - 3$: $x - 3 = 8 - 3 = 5$

The original fraction is $\dfrac{5}{8}$.

4. GEOMETRY

Types of Angles

4.1 ANGLES AND LINES

• Angles may be classified according to their degree measures:

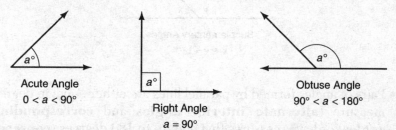

Acute Angle
$0 < a < 90°$

Right Angle
$a = 90°$

Obtuse Angle
$90° < a < 180°$

• Two angles are **supplementary** if their degree measures add up
to 180 and are **complementary** if their degree measures add up
to 90.

- The opposite pairs of angles formed when two lines intersect are called **vertical angles**. Pairs of vertical angles are equal in measure.

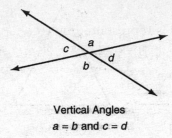

Vertical Angles
$a = b$ and $c = d$

- When the noncommon sides of two adjacent angles form a line, the measures of the two adjacent angles add up to 180 degrees.

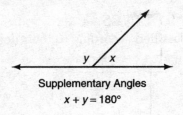

Supplementary Angles
$x + y = 180°$

- Pairs of angles formed by parallel lines are either equal in degree measure (**alternate interior angles** and **corresponding angles**) or have measures that add up to 180 degrees (**consecutive interior angles**).

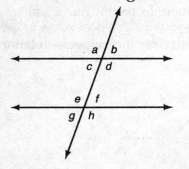

1. Alternate Interior Angles
   $c = f$ and $d = e$
2. Corresponding Angles
   $a = e$ and $c = g$
   $b = f$ and $d = h$
3. Supplementary Interior Angles
   $e + c = 180$ and $d + f = 180$

4.2 TYPES OF TRIANGLES

- Triangles may be classified by the number of sides that have the same length:

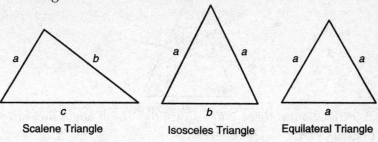

Scalene Triangle   Isosceles Triangle   Equilateral Triangle

- A triangle may also be classified by the degree measure of its largest angle:

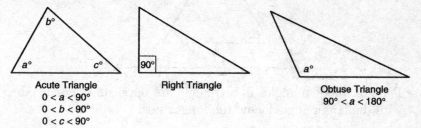

Acute Triangle   Right Triangle   Obtuse Triangle
$0 < a < 90°$ $90° < a < 180°$
$0 < b < 90°$
$0 < c < 90°$

4.3 ANGLES OF A TRIANGLE

- The measures of the three angles of a triangle add up to 180 degrees.
- An exterior angle of a triangle is equal in degree measure to the sum of the degree measures of the two nonadjacent interior angles.

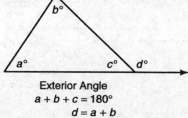

Exterior Angle
$a + b + c = 180°$
$d = a + b$

- If a triangle has two equal sides (angles), then the angles (sides) opposite them are also equal.

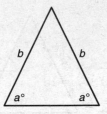

- If a triangle has three equal sides, then each angle measures 60 degrees.

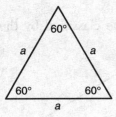

- Pairs of unequal angles of a triangle are opposite unequal sides with the larger angle facing the longer side.

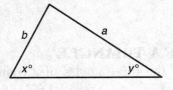

If $x > y$, then $a > b$.
If $a > b$, then $x > y$.

- Each side of a triangle must be shorter than the sum of the lengths of the other two sides and longer than their difference. For example, if the lengths of the two sides of a triangle are 9 and 4, the length of the third side, x, must be shorter than $9 + 4 = 13$ and longer than $9 - 4 = 5$.

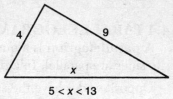

$5 < x < 13$

- For a polygon with n sides, the n interior angles of the polygon add up to

$$(n\ \ 2) \times 180°.$$

Thus, the four angles of a quadrilateral add up to $(4 - 2) \times 180° = 360°$; the five angles of a pentagon add up to $(5 - 2) \times 180° = 3 \times 180° = 540°$; and so forth.

- The sum of the n exterior angles of an n-sided polygon, one angle at each corner, is $360°$.

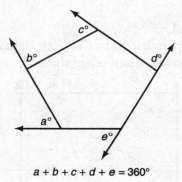

$a + b + c + d + e = 360°$

Special Quadrilaterals

4.4 PARALLELOGRAM

A **parallelogram** is a quadrilateral in which both pairs of opposite sides are parallel. In a parallelogram:

- Opposite sides have the same length.
- Opposite angles have equal measures.
- Diagonals bisect each other.

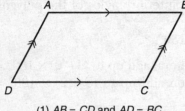

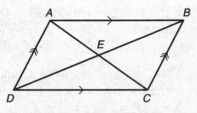

(1) $AB = CD$ and $AD = BC$ (3) $AE = EC$ and $BE = ED$

(2) $\angle B \cong \angle D$ and $\angle A \cong \angle C$

4.5 RECTANGLE

A **rectangle** is a parallelogram with four right angles. The diagonals of a rectangle have the same length.

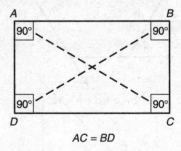

$AC = BD$

4.6 RHOMBUS

A **rhombus** is a parallelogram in which all four sides have the same length. In a rhombus:

- The diagonals intersect at right angles.
- The diagonals bisect the angles at opposite corners of the rhombus.

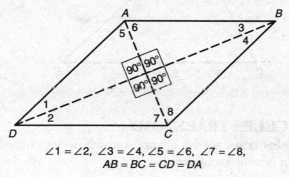

$\angle 1 = \angle 2$, $\angle 3 = \angle 4$, $\angle 5 = \angle 6$, $\angle 7 = \angle 8$,
$AB = BC = CD = DA$

4.7 SQUARE

A **square** is a parallelogram in which all four sides have the same length and all four angles are right angles. Since a square has all of the properties of a rhombus and a rectangle, in a square:

- The diagonals have the same length.
- Each diagonal bisects the two angles at opposite corners of the square.

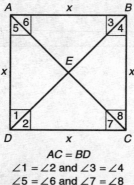

$AC = BD$
$\angle 1 = \angle 2$ and $\angle 3 = \angle 4$
$\angle 5 = \angle 6$ and $\angle 7 = \angle 8$

4.8 TRAPEZOID

A **trapezoid** is a quadrilateral in which one pair of sides are parallel and one pair of sides are not parallel. The parallel sides are called **bases**, and the nonparallel sides are called **legs**.

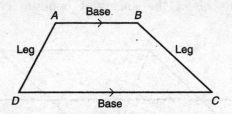

4.9 ISOSCELES TRAPEZOID

An **isosceles trapezoid** is a trapezoid whose legs have the same length. In an isosceles trapezoid:

- The diagonals have the same length.
- The lower base angles are equal in measure and the upper base angles are equal in measure.

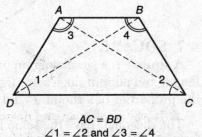

$AC = BD$

$\angle 1 = \angle 2$ and $\angle 3 = \angle 4$

Practice Exercises

1. In the accompanying diagram, $\overleftrightarrow{AB}$ and $\overleftrightarrow{CD}$ intersect at E. If $m\angle AEC = 2x + 40$ and $m\angle CEB = x + 20$, find x.

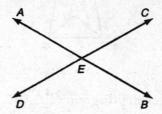

2. In the accompanying diagram, $\overline{AB} \perp \overline{BC}, \overline{DB} \perp \overline{BE}$, $m\angle CBE = x$, $m\angle DBC = y$, and $m\angle ABD = z$. Which statement *must* be true?

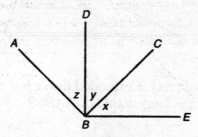

(1) $x = y$ (3) $y = z$
(2) $x = z$ (4) $2y = x + z$

3. The measures of the three angles of a triangle are in the ratio 2:3:4. Find the measure of the largest angle of the triangle.

4. In the accompanying diagram of $\triangle ABC$, $\overline{AB} \cong \overline{AC}$, $\overline{BD}$, and $\overline{DC}$ are angle bisectors, and m$\angle BAC = 20$. Find m$\angle BDC$.

5. In the accompanying diagram of $\triangle ABC$, $\overline{BD}$ is drawn so that $\overline{BD} \cong \overline{DC}$. If m$\angle C = 70$, find m$\angle BDA$.

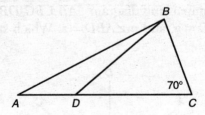

6. In $\triangle ABC$, side $\overline{AC}$ is extended through C to D. If m$\angle DCB = 50$, which is the longest side of $\triangle ABC$?

7. In the accompanying diagram of parallelogram *ABCD,* side $\overline{AD}$ is extended through *D* to *E* and $\overline{DB}$ is a diagonal. If m∠*EDC* = 65 and m∠*CBD* = 85, find m∠*CDB.*

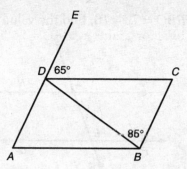

8. In the accompanying diagram, $\overleftrightarrow{AB}$, $\overleftrightarrow{CD}$, $\overleftrightarrow{EF}$, and $\overleftrightarrow{GH}$ are straight lines. If m∠*w* = 30, m∠*x* = 30, and m∠*z* = 120, find m∠*y*.

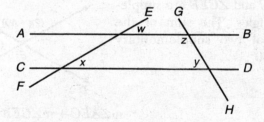

9. In the accompanying diagram transversal $\overleftrightarrow{RS}$ intersects parallel lines $\overleftrightarrow{MN}$ and $\overleftrightarrow{PQ}$ at A and B, respectively. If m$\angle RAN$ = $3x + 24$, m$\angle RBQ = 7x - 16$, find the value of x.

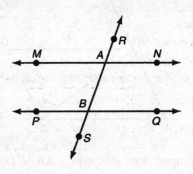

Solutions

1. $\angle AEC$ and $\angle CEB$ are supplementary angles. The sum of the measures of two supplementary angles is 180°:

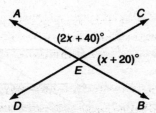

$$\text{m}\angle AEC + \text{m}\angle CEB = 180$$
$$2x + 40 + x + 20 = 180$$
$$3x + 60 = 180$$
$$3x = 180 - 60$$
$$= 120$$
$$x = 40$$

$x = \textbf{40}$.

2. Perpendicular lines meet at right angles:

$\angle ABC$ is a right angle.

$\angle DBE$ is a right angle.

If the sum of two angles equals a right angle, the angles are complementary:

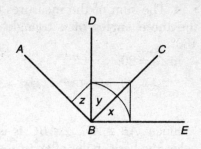

z is the complement of y,

x is the complement of y.

Complements of the same angle are equal in measure: $x = z$.

The correct choice is **(2)**.

3. Let $2x$, $3x$, and $4x$ represent, respectively, the degree measures of the three angles of the triangle.

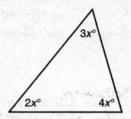

The sum of the measures of the three angles of a triangle is $180°$:

$$2x + 3x + 4x = 80$$
$$9x = 180$$

Divide both sides of the equation by 9:

$$x = 20$$

The measure of the largest angle is $4x$:

$$4x = 4(20) = 80$$

The measure of the largest angle is **80°**.

4. The sum of the measures of the three angles of a triangle is 180°:

m∠a = 20:

$$m\angle A + m\angle B = 180$$
$$20 + m\angle B + m\angle C = 180$$
$$m\angle B + m\angle C = 180 - 20$$
$$= 160$$

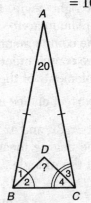

Since $\overline{AB} \cong \overline{AC}$, △ABC is isosceles; the base angles of an isosceles triangle are equal in measure.

$$m\angle B = m\angle C$$
$$2(m\angle B) = 160$$
$$m\angle B = 80$$
$$m\angle C = 80$$

Since $\overline{DB}$ and $\overline{DC}$ are angle bisectors, m∠1 = m∠2, and m∠3 = m∠4:

$$2(m\angle 2) = 80 \quad 2(m\angle 4) = 80$$
$$m\angle 2 = 40 \qquad m\angle 4 = 40$$

In △BDC, the sum of the measures of the three angles is 180°:

$$m\angle BDC + m\angle 2 + m\angle 4 = 180$$
$$m\angle BDC + 40 + 40 = 180$$
$$m\angle BDC + 80 = 180$$
$$m\angle BDC = 180 - 80$$
$$= 100$$

m∠BDC = **100**.

5. Since $\overline{BD} \cong \overline{DC}$, $\triangle BCD$ is isosceles. The base angles of an isosceles triangle are equal in measure:

$$\text{m}\angle DBC = \text{m}\angle C = 70$$

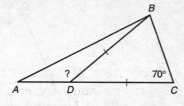

$\angle ADB$ is an exterior angle of $\triangle BDC$. The measure of an exterior angle of a triangle is equal to the sum of the measures of the two remote interior angles:

$$\text{m}\angle BDA = \text{m}\angle DBC + \text{m}\angle C$$
$$\text{m}\angle BDA = 70 + 70$$
$$= 140$$

$\text{m}\angle BDA = \textbf{140}$.

6. If the exterior sides of two adjacent angles form a straight line, they are supplementary: thus $\angle ACB$ and $\angle BCD$ are supplementary.

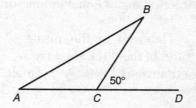

The sum of the measures of two supplementary angles is 180°:

$$\text{m}\angle ACB + \text{m}\angle BCD = 180$$
$$\text{m}\angle ACB + 50 = 180$$
$$\text{m}\angle ACB = 180 - 50$$
$$= 130$$

There can be only one obtuse angle in a triangle. Since $\angle ACB$ is an obtuse angle, it is the largest angle in $\triangle ABC$.

In a triangle, the side opposite the largest angle is the longest side.

Hence, $\overline{AB}$ is the longest side of $\triangle ABC$. (This side can also be designated as AB or c.)

7.

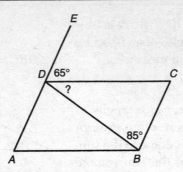

The opposite sides of a parallelogram are parallel:

$$\overline{ADE} \parallel \overline{BC}$$

If two lines are parallel, a transversal makes a pair of alternate
interior angles equal in measure:

$$m\angle EDC = m\angle C$$
$$65 = m\angle C$$

The sum of the mea-
sures of the three angles of
a triangle is 180°:

$$m\angle CDB + m\angle C + m\angle CBD = 180$$
$$m\angle CDB + 65 + 85 = 180$$
$$m\angle CDB + 150 = 180$$
$$m\angle CDB = 180 - 150$$
$$= 30$$

$$m\angle CDB = \mathbf{30}.$$

8.

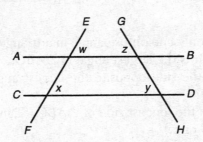

Since $m\angle w = 30$ and $m\angle x = 30$:

$$m\angle w = m\angle x$$

If a transversal to two lines makes a
pair of corresponding angles equal in
measure, the lines are parallel:

$$\overline{AB} \parallel \overline{CD}$$

If two lines are parallel, the interior angles on the same side of a transversal are supplementary, so $\angle y$ is the supplement of $\angle z$.

The sum of the measures of two supplementary angles is $180°$:

It is given that $m\angle z = 120$:

$$m\angle y + m\angle z = 180$$
$$m\angle y + 120 = 180$$
$$m\angle y = 180 - 120$$
$$= 60$$

$m\angle y = \mathbf{60}$.

9.

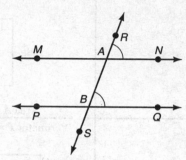

Angles RBQ and RAN are corresponding angles. If two lines are parallel, then corresponding angles have the same measure:

On each side add 16 and subtract $3x$:

$$m\angle RBQ = m\angle RAN$$
$$7x - 16 = 3x + 24$$
$$\underline{-3x + 16 = -3x + 16}$$
$$4x = 40$$

Divide each side of the resulting equation by 4:

$$\frac{4x}{4} = \frac{40}{4}$$
$$x = 10$$

The value of x is **10**.

5. MEASUREMENT

5.1 PERIMETER AND AREA FORMULAS

The distance around a circle is called its **perimeter**. The **area** of a figure is the number of 1×1 squares that it can enclose. Perimeter and area formulas for six types of figures are given in the accompanying table.

Type of Figure	Formulas
1. Square	Perimeter $P = 4 \times$ side length Area $A =$ side $\times$ side
2. Rectangle	Perimeter $P = 2(\text{length} + \text{width})$ Area $A =$ length $\times$ width
3. Parallelogram and triangle	 Area A of $\square ABCD =$ base $\times$ height Area A of $\triangle ABD$ $\frac{1}{2} =$ base $\times$ height

Type of Figure	Formulas
4. Rhombus or square	Area $A = \frac{1}{2} \times \text{diagonal}_1 \times \text{diagonal}_2$
5. Trapezoid	Area $A = \frac{1}{2} \times \text{height} \times (b_1 + b_2)$
6. Circle	Circumference $C = 2 \times \pi \times \text{radius}$ $= \pi \times \text{diameter}$ Area $A = \pi \times (\text{radius})^2$

5.2 FINDING AREAS INDIRECTLY

- The area A of the ring between two circles that have the same center is given by the formula

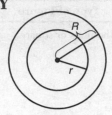

$$A = \pi \times (R)^2 - \pi \times (r)^2$$

- To find the area of a shaded region, try subtracting the areas of the figures that overlap to form that region. Suppose that, in the accompanying figure, each circle has an area of 25π.

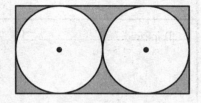

Since $25\pi = \pi r^2 = \pi 5^2$, the radius of each circle is 5, so the diameter is 10. The width of the rectangle is one diameter or 10, and the length of the rectangle is two diameters or 20. Hence, the area of the rectangle is $20 \times 10 = 200$. The area of the shaded region is equal to the difference between the area of the rectangle and the sum of the areas of the two circles, which is $200 - 50\pi$.

5.3 VOLUME FORMULAS

The **volume** of a figure is the number of $1 \times 1 \times 1$ cubes that it can enclose. Volume formulas for three types of solid figures are given in the accompanying table.

Type of Solid Figure	Volume Formula
1. Rectangular box	Volume = length $\times$ width $\times$ height

Type of Solid Figure	Volume Formula
2. Cube	Volume = (side)³ Side
3. Circular cylinder	Volume = area of circular base × height h

5.4 CONGRUENT VERSUS SIMILAR FIGURES

Congruent figures have the same size and the same shape. **Similar figures** have the same shape but may have different sizes.

• The symbol for congruence is ≅.

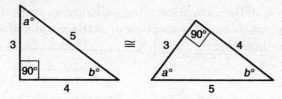

• The symbol for similarity is ~.

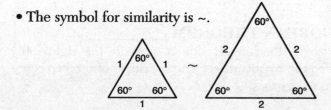

5.5 DETERMINING WHEN TRIANGLES ARE CONGRUENT

Two triangles are congruent when any one of the following conditions is true:

- The three sides of one triangle are congruent to the corresponding three sides of the other triangle ($SSS \cong SSS$).
- Two sides and their included angle of one triangle are congruent to the corresponding parts of the other triangle ($SAS \cong SAS$).
- Two angles and their included side of one triangle are congruent to the corresponding parts of the other triangle ($ASA \cong ASA$).
- Two angles and the side opposite one of these angles of one triangle are congruent to the corresponding parts of the other triangle ($AAS \cong AAS$).
- The hypotenuse and a leg of one right triangle are congruent to the corresponding parts of another right triangle ($HL \cong HL$).

When two triangles are congruent, you may conclude that any pair of corresponding parts of the two figures are congruent.

5.6 DETERMINING WHEN TRIANGLES ARE SIMILAR

If two angles of one triangle have the same degree measures as the corresponding angles of a second triangle, the two triangles are similar. When two polygons are similar, you may conclude that:

- The lengths of their corresponding sides are in proportion.
- Their perimeters have the same ratio as the lengths of any pair of corresponding sides.

Right-Triangle Relationships

5.7 PYTHAGOREAN THEOREM

In a right triangle, the longest side is opposite the right (90°) angle and is called the **hypotenuse**. The measures of the two acute angles add up to 90 degrees.

PYTHAGOREAN THEOREM: $(\text{Side I})^2 + (\text{side II})^2 = (\text{hypotenuse})^2$

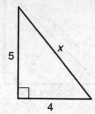

$$4^2 + 5^2 = x^2$$
$$16 + 25 = x^2$$
$$41 = x^2$$
$$x = \sqrt{41}$$

It is not necessary to use the Pythagorean theorem if the lengths of two sides of a right triangle fit one of the following patterns, where n is a positive integer:

1. $\text{Leg}_1 : \text{leg}_2 : \text{hypotenuse} = 3n{:}4n{:}5n$
2. $\text{Leg}_1 : \text{leg}_2 : \text{hypotenuse} = 5n{:}12n{:}13n$

A set of three positive integers that satisfy the Pythagorean relationship is called a Pythagorean triple. For example, if the length and width of a rectangle are 8 and 6, respectively, you don't need the Pythagorean theorem to find the diagonal (hypotenuse). Since $3{:}4{:}5 = 6{:}8{:}10$ (where $n = 2$), the length of a diagonal is 10.

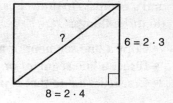

5.8 RIGHT-TRIANGLE TRIGONOMETRY

The trigonometric ratios sine, cosine, and tangent allow you to find the measure of a part of a right triangle indirectly when you know the measures of another side and an angle of the triangle.

$\text{Sin } A = \dfrac{\text{Opposite}}{\text{Hypotenuse}}$	$\text{Cos } A = \dfrac{\text{Adjacent}}{\text{Hypotenuse}}$	$\text{Tan } A = \dfrac{\text{Opposite}}{\text{Adjacent}}$
S O H	C A H	T O A

Notice that the first letters of the three key words in the definitions of the three trigonometric ratios spell the word *SOHCAHTOA*. Remembering *SOHCAHTOA* can help you recall the definitions of sine, cosine, and tangent.

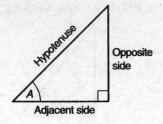

Constructions

5.9

A geometric construction requires a straightedge to draw lines and a compass to draw arcs. Using these tools, you should be able to do three things:

- Copy a line segment or an angle.
- Bisect a line segment or an angle.
- Construct a line that is perpendicular or parallel to another line.

Locus

5.10 BASIC LOCI

You can think of a *locus* (plural, *loci*) as a description of a path that satisfies one or more conditions. A **locus** is defined as the set of points, and only those points, that satisfy a given set of conditions. To help determine a locus, draw a diagram.

- The locus of points at a fixed distance of *d* units from a point *P* is a circle with center at *P* and radius of length *d*.

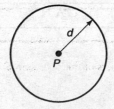

The locus consists of the points on the circumference of the circle.

- The locus of points at a fixed distance of d units from a given line is a pair of parallel lines with each line d units from the given line.

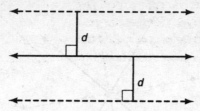

The locus consists of the points on either of the dotted lines.

- The locus of points equidistant from two fixed points is the perpendicular bisector of the line segment whose endpoints are the fixed points.

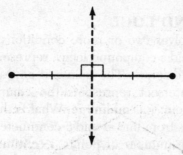

The locus consists of the points on the dotted line.

- The locus of points equidistant from two parallel lines is the parallel line midway between them.

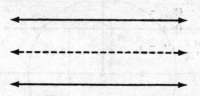

The locus consists of the points on the dotted line.

• The locus of points equidistant from two intersecting lines is the bisector of each pair of vertical angles.

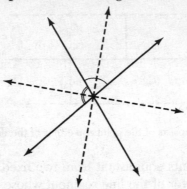

The locus consists of the points on the dotted lines.

5.11 COMPOUND LOCI

A locus that involves two or more conditions is called a **compound locus**. To find a compound locus, represent each locus condition on the same diagram. The points, if any, at which all of the locus conditions intersect represents the compound locus. For example, suppose point P is on line m. What is the total number of points 3 centimeters from line m and 5 centimeters from point P?

To determine the number of points 3 centimeters from line m [condition 1] and 5 centimeters from point P [condition 2], represent the two locus conditions as shown in the accompanying diagram and count the number of points at which the loci intersect.

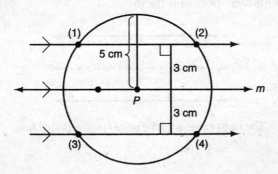

- The locus of points 3 centimeters from line m is a pair of lines each parallel to line m and each 3 centimeters from line m.
- The locus of points 5 centimeters from point P is a circle with center at P and a radius length of 5 centimeters.

Since the radius of the circle is greater than 3, the circle intersects each of the parallel lines at two points. Therefore, the two locus conditions intersect at a total of four points.

The total number of points that satisfy the given conditions is 4.

Practice Exercises

1. The ratio of the corresponding sides of two similar polygons is 2:3. If the perimeter of the larger polygon is 27, find the perimeter of the smaller polygon.

2. Three sides of a triangle measure 4, 5, and 8. Find the length of the longest side of a similar triangle whose perimeter is 51.

3. In a rectangle, the length is twice the width, and the perimeter is 48. Find the area of the rectangle.

4. In the accompanying diagram, $ABCD$ is a rectangle. Diameter MN of circle O is perpendicular to $\overline{BC}$ at M and to $\overline{AD}$ at N, $AD = 8$, and $CD = 6$. (Answers may be left in terms of π.)

   a. What is the perimeter of rectangle $ABCD$?

   b. What is the circumference of circle O?

   c. What is the area of rectangle $ABCD$?

   d. What is the area of circle O?

   e. What is the area of the shaded region of the diagram?

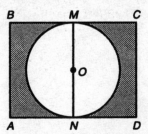

5. The locus of the midpoints of the radii of a circle is
    (1) a point   (3) a line
    (2) two lines   (4) a circle

6. How many points are equidistant from two intersecting lines, ℓ and m, and 3 units from the point of intersection of the lines?
    (1) 1   (3) 3
    (2) 2   (4) 4

7. To locate a point equidistant from the vertices of a triangle, construct
    (1) the perpendicular bisectors of the sides
    (2) the angle bisectors
    (3) the altitudes
    (4) the medians

8. In the accompanying diagram of rectangle $ABCD$, $AC = 22$ and m$\angle CAB = 24$.
  a. Find AB to the *nearest integer*.
  b. Find BC to the *nearest integer*.
  c. Using the results from parts *a* and *b*, find the number of square units in the area of $ABCD$.

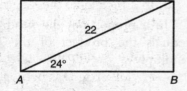

9. In the accompanying diagram, $\triangle IHJ \sim \triangle LKJ$. If $IH = 5$, $HJ = 2$, and $LK = 7$, find KJ.

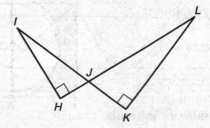

10. In the accompanying diagram, $ABCD$ is an isosceles trapezoid, $AD = BC = 5$, $AB = 10$, and $DC = 18$. Find the length of altitude $\overline{AE}$.

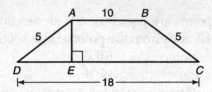

Solutions

1. Let p = perimeter of the smaller polygon.

Let P = perimeter of the larger polygon.

The perimeters of two similar polygons have the same ratio as any two corresponding sides: $\dfrac{p}{P} = \dfrac{2}{3}$

Since $P = 27$: $\dfrac{p}{27} = \dfrac{2}{3}$

In a proportion, the product of the means equals the product of the extremes (cross-multiply):

$$3p = 2(27)$$
$$= 54$$
$$p = 18$$

The perimeter of the smaller polygon is **18**.

2.

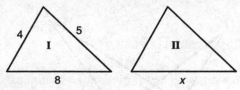

Let P = perimeter of triangle I.

Let x = longest side of the similar triangle, II.

Then x will correspond to 8, the longest side of triangle I.

The perimeter of a triangle is the sum of the lengths of the three sides:

$$P = 4 + 5 + 8$$
$$= 17$$

The perimeters of two similar triangles have the same ratio as any two corresponding sides:

$$\frac{17}{51} = \frac{8}{x}$$

Reduce the fraction on the left by dividing its numerator and denominator by 17:

$$\frac{1}{3} = \frac{8}{x}$$

In a proportion, the product of the means is equal to the product of the extremes (cross-multiply):

$$x = 3(8)$$
$$= 24$$

The longest side is **24**.

3. Let x = width of the rectangle. Then $2x$ = length of the rectangle.

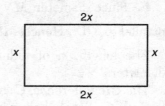

The perimeter, P, equals the sum of the lengths of all of the sides:

$$P = x + 2x + x + 2x$$

The perimeter is 48:

$$48 = 6x$$
$$8 = x \text{ (width is 8)}$$
$$2x = 2(8) = 16 \text{ (length is 16)}$$

The area, A, of a rectangle is equal to the product of its length and width:

$$A = 8(16)$$
$$= 128$$

The area of the rectangle is **128**.

4. a. The opposite sides of a rectangle are equal:

$$BC = AD = 8; \quad AB = CD = 6$$

The perimeter, P, of a rectangle equals the sum of the lengths of all four sides:

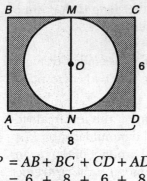

$$P = AB + BC + CD + AD$$
$$= 6 + 8 + 6 + 8$$
$$= 28$$

The perimeter of $ABCD$ is **28**.

b. Since diameter $\overline{MN}$ is perpendicular to $\overline{BC}$ and $\overline{AD}$, it is parallel to $\overline{CD}$. Hence, $MN = CD = 6$.

The length, r, of radius $\overline{MO}$ is one-half the diameter: $r = 3$

The circumference, C, of a circle is given by the formula $C = 2\pi r$, where r is the length of the radius: $C = 2\pi(3)$

The circumference of circle O is **6π**. $= 6\pi$

c. The area of a rectangle is equal to the product of its length and width: Area of $ABCD = (8)(6)$
$$= 48$$

The area of $ABCD$ is **48** square units.

d. The area, A, of a circle is given by the formula $A = \pi r^2$, where r is the radius: $A = \pi(3)^2$
$$= 9\pi$$

The area of the circle is **9π**.

e. The area of the shaded region is equal to the area of the rectangle minus the area of the circle: $48 - 9\pi$

The area of the shaded region is **48 square units − 9π**.

5. If the outer circle is the original circle, the locus of the midpoints of its radii is a smaller concentric circle whose radius is exactly half that of the original.

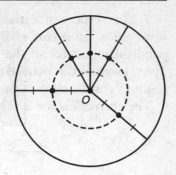

The correct choice is **(4)**.

6. All points equidistant from the intersecting lines, ℓ and m, are on the angle bisectors of the angles formed at the intersection of ℓ and m. The bisectors are shown as dashed lines in the drawing.

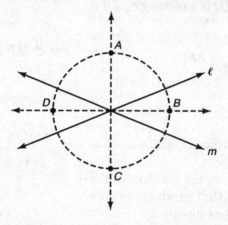

All points 3 units from the intersection of ℓ and m are on the circle whose center is the intersection and whose radius is 3 units.

Points satisfying both conditions are at the intersection of the circle with the angle bisectors. There are 4 such points, labeled A, B, C, and D in the accompanying figure.

The correct choice is **(4)**.

7. The vertices of a triangle are really the two endpoints of each of its sides. All points equidistant from two points lie on the perpendicular bisector of the line segment joining those two points. Thus, locating the point equidistant from the vertices of a triangle requires construction of the perpendicular bisectors of the sides.

The correct choice is **(1)**.

8.

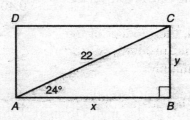

a. Since *ABCD* is a rectangle, ∠*B* is a right angle.

Let $x = AB$.

In right triangle *ABC*:

$$\cos \angle CAB = \frac{\text{adjacent leg}}{\text{hypotenuse}}$$

$$\cos 24° = \frac{x}{22}$$

Using a scientific calculator, cos 24° = 0.9135:

$$\frac{0.9135}{1} = \frac{x}{22}$$

In a proportion, the product of the means equals the product of the extremes (cross-multiply):

$$x = 22(0.9135)$$
$$= 20.0970$$

Round off to the nearest integer:
AB = **20** to the *nearest integer*.

$$= 20$$

b. Let $y = BC$.

In right triangle ABC:

$$\sin \angle CAB = \frac{\text{opposite leg}}{\text{hypotenuse}}$$

$$\sin 24° = \frac{y}{22}$$

Using a scientific calculator, $\sin 24°$ = 0.4067:

$$\frac{0.4067}{1} = \frac{y}{22}$$

In a proportion, the product of the means equals the product of the extremes (cross-multiply):

$$y = 22(0.4067)$$
$$= 8.9474$$

Round off to the nearest integer: $BC = $ **9** to the *nearest integer*.

$$= 9$$

c. The are of a rectangle is equal to the product of its length and width:

$$\text{Area of } ABCD = (AB)(BC)$$
$$= (20(9)$$
$$= 180$$

The area of $ABCD$ is **180** square units.

9. Lengths of corresponding sides of similar triangles are in proportion. Hence:

$$\frac{KJ}{LK} = \frac{HJ}{IH}$$

Since $IH = 5$, $HJ = 2$, and $LK = 7$:

$$\frac{KJ}{7} = \frac{2}{5}$$

In proportion the product of the means equals the product of the extremes?

$$(5)(KJ) = (7)(2)$$

Divide each side of the equation by 5:

The length of $\overline{KJ}$ is $\dfrac{14}{5}$.

$$KJ = \frac{14}{5}$$

10.

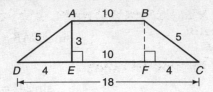

In isosceles trapezoid $ABCD$, draw an altitude from B to side CD, intersecting CD at F. Since $ABFE$ is a rectangle, $AB = EF = 10$.

Right triangles AED and BFC are congruent by the hypotenuse-leg (HL) method, so $DE = FC = \dfrac{1}{2}(8) = 4$.

The lengths of the sides of right triangle AED from a $3 - 4 - 5$ Pythagorean triple in which $AE = 3$.

The length of altitude $\overline{AE}$ is **3**.

6. SYMMETRY, TRANSFORMATIONS, AND GRAPHING

Symmetry and Transformations

6.1 TYPES OF SYMMETRY

• A figure has **line symmetry** if a line can be drawn that separates the figure into two parts in such a way that, if the figure is folded along the line, the two parts will coincide, as shown in the accompanying diagrams.

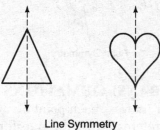

Line Symmetry

A figure may have more than one line of symmetry. For example, a rectangle has vertical and horizontal lines of symmetry, as shown in the accompanying diagram.

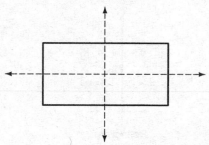

Vertical and Horizontal Line Symmetry

• A figure has **point symmetry** if there exists a point X such that, if any line is drawn through X and intersects the figure in a point

P, it will also intersect the figure in another point P', with $PX = XP'$. The letter **S** has point symmetry, as shown in the accompanying diagram. When a figure has point symmetry about a point X, turning the figure 180 degrees about point X does not change how the figure looks.

Point Symmetry

6.2 TYPES OF TRANSFORMATIONS

A **transformation** "moves" each point of a figure according to a stated rule. The figure created by the transformation is called the **image** of the original figure.

• A **reflection** is a transformation that "flips" a figure over the line of reflection, as shown in the accompanying diagram, so that each figure is the mirror image of the other.

Reflection

- A **translation** is a transformation that "slides" a figure up or down, slides it sideways, or does both, as shown in the accompanying diagram.

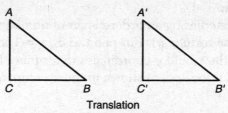

Translation

- A **rotation** is a transformation that "turns" a figure with respect to a fixed reference point, as shown in the accompanying diagram.

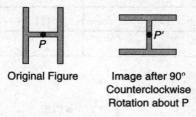

Original Figure   Image after 90°
Counterclockwise
Rotation about P

- Reflections, translations, and rotations do not change the size of a figure. A **dilation** is a transformation that changes the size but not the shape of a figure, as shown in the dilation of the letter **M** accompanying diagram.

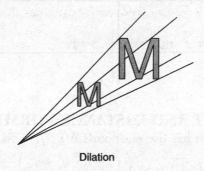

Dilation

6.3 THE COORDINATE PLANE

A coordinate system can be formed by drawing horizontal and vertical number lines that intersect at their common zero point, called the **origin**.

- Points are located using an ordered pair of numbers, called **coordinates**, of the form (x,y). For point $(3,5)$, $x = 3$ and $y = 5$.
- The signs of the x- and y-coordinates determine the quadrant in which a point is located, as shown in the accompanying table and diagram.

(Sign of x, sign of y)	Quadrant
$(+, +)$	I
$(-, +)$	II
$(-, -)$	III
$(+, -)$	IV

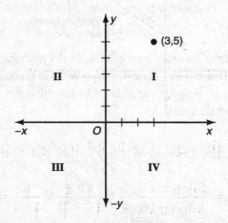

6.4 MIDPOINT AND DISTANCE FORMULAS

If a line segment has the endpoints $A(x_A, y_A)$ and $B(x_B, y_B)$, then:

- The midpoint of $\overline{AB}$ is at $\left(\dfrac{x_A + x_B}{2}, \dfrac{y_a + y_B}{2} \right)$.

- The length of $\overline{AB}$ is $\sqrt{\left(x_B - x_A\right)^2 + \left(y_B - y_A\right)^2}$. To find the distance between points A and B, find the length of $\overline{AB}$.

6.5 SLOPE OF A LINE

The **slope** of a line is a number that represents the steepness of the line.

- The slope of the nonvertical line that passes through $A(x_A, y_A)$ and $B(x_B, y_B)$ in the accompanying diagram is given by the formula

$$\text{Slope} = \frac{\text{vertical change (difference) in } y}{\text{horizontal change (difference) in } x} = \frac{y_B - y_A}{x_B - x_A}$$

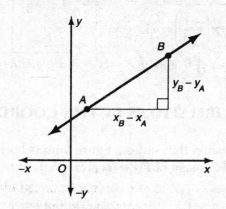

For example, the slope m of the line that passes through $A(-1,2)$ and $B(1,8)$ is

$$m = \frac{\text{difference in } y}{\text{difference in } x} = \frac{8-2}{1-(-1)} = \frac{6}{1+1} = 3$$

- The slope of a horizontal line is 0, and the slope of a vertical line is not defined.
- A line that rises from left to right has a positive slope.
- A line that falls from left to right has a negative slope.

6.6 EQUATION OF A LINE

The set of points (x,y) on a nonvertical line satisfy the equation $y = mx + b$, where m represents the slope of the line and b is the y-coordinate of the point at which the line crosses the y-axis. When the equations of two different lines are both written in the form $y = mx + b$, comparing the values of m can tell you whether the lines are parallel, perpendicular, or neither parallel nor perpendicular.

- If two lines have the same slope, the lines are *parallel*. The lines $y = 3x - 7$ and $y = 3x + 1$ are parallel since the value of m, the coefficient of x, is the same.

- If two lines have slopes whose product is –1, the lines are *perpendicular*. The lines $y = 2x + 3$ and $y = -\dfrac{1}{2}x + 5$ are perpendicular since $(2) \times \left(-\dfrac{1}{2}\right) = -1$.

- The lines $y = x + 6$ and $y = -2x + 3$ are *neither parallel nor perpendicular.*

6.7 TRANSFORMATIONS IN THE COORDINATE PLANE

Under a translation that slides a figure h units horizontally and k units vertically, the image of $P(x,y)$ is $P'(x + h, y + k)$.

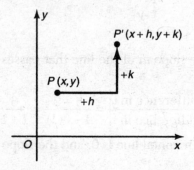

6.8 REFLECTIONS IN THE COORDINATE PLANE

Points may be reflected in the coordinate axis, in the origin, or in the line $y = x$ according to the following rules:

- In a reflection in the x-axis, a point $P(x,y)$ is flipped over the x-axis, so its image is $P'(x,-y)$.
- In a reflection in the y-axis, a point $P(x,y)$ is flipped over the y-axis, so its image is $P''(-x,y)$.

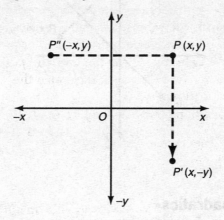

- In a reflection in the origin, the image of a point $R(x,y)$ is the point located diagonally across the origin and the same distance from it as R. Thus, the image of $R(x,y)$ is $R'(-x,-y)$.

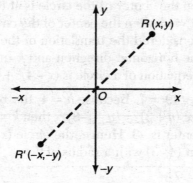

- In a reflection in the line $y = x$, the image of a point $P(x,y)$ is $P'(y,x)$.

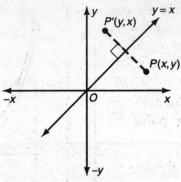

Graphs of Quadratics

6.9 CIRCLES

If r represents the length of a radius of a circle, then an equation of the circle is:

- $x^2 + y^2 = r^2$ when the center of the circle is at $(0,0)$.
- $(x - h)^2 + (y - k)^2 = r^2$ when the center of the circle is at (h,k). This circle may be considered the translation of the circle $x^2 + y^2 = r^2$ by h units in the horizontal direction and k units in the vertical direction. If an equation of a circle is $(x - 4)^2 + (y + 3)^2 = 49$, then $r^2 = 49$, so $r = \sqrt{49} = 7$. Because $h = 4$, the x-coordinate of the center is 4. Since $(y + 3)^2 = (y - [-3])^2$, then $k = -3$, so the y-coordinate of the center is -3. Hence, the circle $(x - 4)^2 + (y + 3)^2 = 49$ is centered at $(4,-3)$ with a radius of 7.

6.10 PARABOLAS

The graph of the quadratic equation $y = ax^2 + bx + c$ is a U-shaped curve, called a **parabola**, that has a vertical axis of symmetry. The point at which the axis of symmetry intersects the parabola is called the **turning point** or **vertex**.

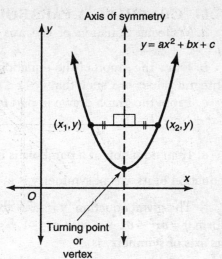

- An equation of the axis of symmetry is $x = -\dfrac{b}{2a}$.

- Since the x-coordinate of the turning point is $-\dfrac{b}{2a}$, substituting this value for x in $y = ax^2 + bx + c$ gives the y-coordinate of the turning point.
- The sign of a, the coefficient of the x^2-term, determines whether the turning point of a parabola is a minimum or a maximum point on the graph. If $a > 0$, the turning point is the *lowest* point on the parabola. If $a < 0$, the turning point is the *highest* point on the parabola.

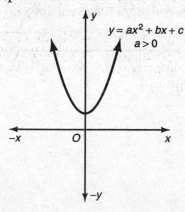

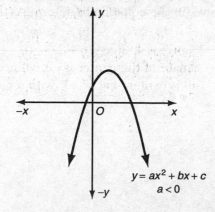

6.11 GRAPHING A PARABOLA: AN EXAMPLE

a. Write an equation of the axis of symmetry of the graph of $y = -x^2 + 8x - 7$.

b. Draw the graph of the equation $y = -x^2 + 8x - 7$, including all integral values of x such that $0 \le x \le 8$.

c. From the graph drawn in part *b*, find the roots of the equation $-x^2 + 8x - 7 = 0$.

a. If an equation of a parabola is in the form $y = ax^2 + bx + c$, an equation of its axis of symmetry is in the form $x = \dfrac{-b}{2a}$.

The given equation, $y = -x^2 + 8x - 7$, is in the form $y = ax^2 + bx + c$, with $a = -1$, $b = 8$, and $c = -7$. An equation of its axis of symmetry is

$$x = \frac{-8}{2(-1)}$$
$$= \frac{-8}{-2}$$
$$= 4$$

An equation of the axis of symmetry is $x = 4$.

b. To graph $y = -x^2 + 8x - 7$ for $0 \le x \le 8$, prepare a table of values for x and y by substituting each integral value of x, from 0 to 8 inclusive, in the equation to determine the corresponding value of y.

x	$-x^2 + 8x - 7$	$= y$
0	$-0^2 + 8(0) - 7 = 0 + 0 - 7$	$= -7$
1	$-1^2 + 8(1) - 7 = -1 + 8 - 7$	$= 0$
2	$-2^2 + 8(2) - 7 = -4 + 16 - 7$	$= 5$
3	$-3^2 + 8(3) - 7 = -9 + 24 - 7$	$= 8$
4	$-4^2 + 8(4) - 7 = -16 + 32 - 7$	$= 9$
5	$-5^2 + 8(5) - 7 = -25 + 40 - 7$	$= 8$
6	$-6^2 + 8(6) - 7 = -36 + 48 - 7$	$= 5$
7	$-7^2 + 8(7) - 7 = -49 + 56 - 7$	$= 0$
8	$-8^2 + 8(8) - 7 = -64 + 64 - 7$	$= -7$

Plot points (0,–7) (1,0), (2,5), (3,8), (4,9), (5,8), (6,5), (7,0), and (8,–7), and draw a smooth curve through them. This curve is the graph of $y = -x^2 + 8x - 7$ for $0 \leq x \leq 8$.

c. The roots of $-x^2 + 8x - 7 = 0$ are the values of x on the graph at the points where $y = 0$, which is the x-axis. The graph crosses the x-axis at two points: $x = 1$ and $x = 7$.

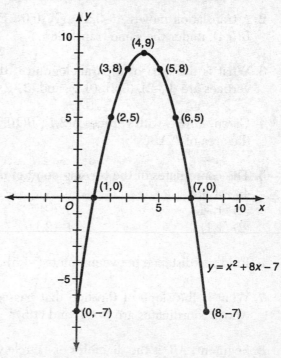

The roots are **1** and **7**.

Practice Exercises

1. Find A', the image of $A(3,5)$, after a reflection in the line $y = x$.

2. A translation moves $A(-3,2)$ to $A'(0,0)$. Find B', the image of $B(5,4)$, under the same translation.

3. What is the area of a parallelogram if the coordinates of its vertices are $(0,-2)$, $(3,2)$, $(8,2)$, and $(5,-2)$?

4. Given: $\triangle ABC$ with vertices $A(2,1)$, $B(10,7)$, and $C(4,10)$. Find the area of $\triangle ABC$.

5. The coordinates of the turning point of the graph of $y = 2x^2 - 4x + 1$ are
   (1) $(1,-1)$ (3) $(-1,5)$
   (2) $(1,1)$ (4) $(2,1)$

6. Find the distance between points $(-1,5)$ and $(-7,3)$.

7. What is the slope of the line that passes through the points whose coordinates are $(-1,4)$ and $(1,5)$?

8. Segment $\overline{AB}$ is the diameter of a circle whose center is $(2,0)$. If the coordinates of A are $(0,-3)$, find the coordinates of B.

9. Which is an equation of the circle whose center is (1,3) and whose radius is 2?
(1) $(x - 1)^2 + (y - 3)^2 = 2$
(2) $(x - 1)^2 + (y - 3)^2 = 4$
(3) $x^2 + y^2 = 4$
(4) $(x + 1)^2 + (y + 3)^2 = 4$

10. A point on the graph of $x + 3y = 13$ is
(1) (4,4)   (2) (–2,3)   (3) (–5,6)   (4) (4,–3)

11. What is the slope of the line whose equation is $4y = 3x + 16$?

12. The graph of $y = 3x - 4$ is parallel to the graph of
(1) $y = 4x - 3$ (3) $y = -3x + 4$
(2) $y = 3x + 4$ (4) $y = 3$

13. The equation of a line whose slope is 2 and whose y-intercept is –2 is
(1) $2y = x - 2$ (3) $y = -2x + 2$
(2) $y = -2$ (4) $y = 2x - 2$

Solutions

1. The line $y = x$ is a line through the origin inclined at an angle of $45°$ to the positive directions of the x- and y-axes.

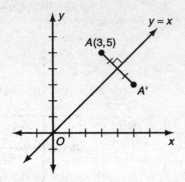

The image of a point $P(x,y)$ after a reflection in the line $y = x$ is $P'(y,x)$, so the image of $A(3,5)$, is $A'(5,3)$.

The image is **$A'(5,3)$**.

2.

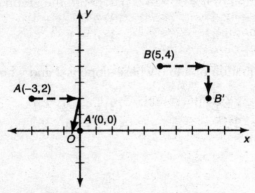

The translation that moves $A(-3,2)$ to $A'(0,0)$ is $T(x + 3, y - 2)$ since the point is moved 3 units to the right and 2 units down.

At $B(5,4)$, $x = 5$ and $y = 4$. After the translation, $B(5,4)$ becomes $B'(5 + 3, 4 - 2)$ or $B'(8,2)$.

The image is **$B'(8,2)$**.

3. Drop altitude $\overline{BE}$ perpendicular to $\overline{AD}$. The coordinates of E are $(3,-2)$.

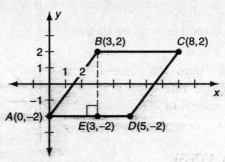

The area of a parallelogram equals the product of the length of its base and its altitude:

Since $AD = 5 - 0 = 5$ and $BE = 2 - (-2) = 2 + 2 = 4$:

Area of $\square ABCD = AD \times BE$

$= 5 \quad \times 4$

$= 20$

The area is **20** square units.

4. Drop perpendiculars $\overline{AD}$, $\overline{CE}$, and $\overline{BF}$ to the x-axis.

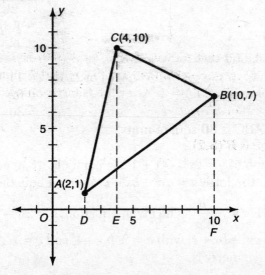

Area of $\triangle ABC$ = area of trapezoid $DACE$ + area of trapezoid $ECBF$ − area of trapezoid $DABF$.

The area, A, of a trapezoid whose altitude is h and the length of whose bases are b_1 and b_2 is given by the formula $A = \dfrac{1}{2} h(b_1 + b_2)$.

<u>Trapezoid $DACE$:</u>
$$h = DE = 2, b_1 = DA = 1, b_2 = EC = 10:$$
$$A = \frac{1}{2} (2)(1 + 10)$$
$$= 1(11)$$
$$= 11$$

<u>Trapezoid $ECBF$:</u>
$$h = EF = 6, b_1 = FB = 7, b_2 = EC = 10:$$
$$A = \frac{1}{2} (6)(7 + 10)$$
$$= 3(17)$$
$$= 51$$

<u>Trapezoid $DABF$:</u>
$$h = DF = 8, b_1 = DA = 1, b_2 = FB = 7:$$
$$A = \frac{1}{2} (8)(1 + 7)$$
$$= 4(8)$$
$$= 32$$

$$\text{Area of } \triangle ABC = 11 + 51 - 32$$
$$= 62 \qquad - 32$$
$$= 30$$

Area of $\triangle ABC$ is **30** square units.

5. The graph of $y = 2x^2 - 4x + 1$ is a parabola. If an equation of a parabola is in the form $y = ax^2 + bx + c$, then an equation of its axis of symmetry is $x = -\dfrac{b}{2a}$. The given equation, $y = 2x^2 - 4x + 1$, is in the form $y = ax^2 + bx + c$, with $a = 2$, $b = -4$, and $c = 1$. An equation of its axis of symmetry is

$$x = -\frac{-4}{2(2)}$$

$$= -\frac{-4}{4}$$

$$= 1.$$

The axis of symmetry of a parabola passes through its turning point. Therefore the x-coordinate of the turning point in this case is 1. Since the turning point is on the parabola, its coordinates must satisfy the equation of the parabola; substitute 1 for x in the equation to determine the y coordinate of the turning point:

$$
\begin{aligned}
y &= 2(1)^2 - 4(1) + 1 \\
&= 2(1) - 4 + 1 \\
&= 2 - 4 + 1 \\
&= -1
\end{aligned}
$$

The coordinates of the turning point are (1,–1).

The correct choice is **(1)**.

6.

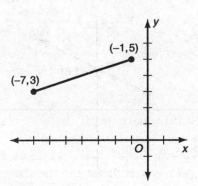

Use the formula for the distance, d, between two points (x_1,y_1) and (x_2,y_2):

$$d = \sqrt{\left(x_2 - x_1\right)^2 + \left(y_2 - y_1\right)^2}$$

Since $x_1 = -1$, $y_1 = 5$, and $x_2 = -7$, $y_2 = 3$:

$$= \sqrt{(-7 - [-1])^2 + (3 - 5)^2}$$

$$= \sqrt{(-7 + 1)^2 \quad + (-2)^2}$$

$$= \sqrt{(-6)^2 \quad\quad + 4}$$

$$= \sqrt{36 + 4}$$

$$= \sqrt{40}$$

A radical expression may be simplified by factoring out any perfect square factor in the radicand:

$$d = \sqrt{4(10)}$$

Remove the perfect square factor from under the radical sign by taking its square root and writing it as a coefficient of the radical:

$$= 2\sqrt{10}$$

The distance is $2\sqrt{10}$.

7.

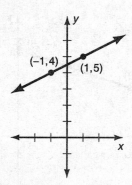

Use the formula for the slope, m, of a line joining points (x_1, y_1) and (x_2, y_2):

$$m = \frac{y_2 - y_1}{x_2 - x_1}$$

Since $x_1 = -1$, $y_1 = 4$, and $x_2 = 1$, $y_2 = 5$:

$$= \frac{5-4}{1-(-1)}$$

$$= \frac{1}{1+1}$$

$$= \frac{1}{2}$$

The slope $= \dfrac{1}{2}$.

8. Since $\overline{AB}$ is a diameter of the circle, the center of the circle, O, is the midpoint of $\overline{AB}$.

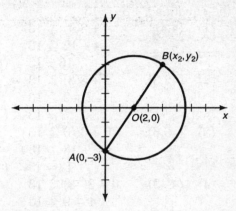

Use the formulas for the coordinates $(\bar{x}, \bar{y})$ of the midpoint of the line segment joining (x_1, y_1) and (x_2, y_2):

$$\bar{x} = \frac{x_1 + x_2}{2} \qquad \bar{y} = \frac{y_1 + y_2}{2}$$

Hence, $\bar{x} = 2$ and $\bar{y} = 0$:

$$2 = \frac{0 + x_2}{2} \qquad 0 = \frac{-3 + y_2}{2}$$

Since $x_1 = 0$, $y_1 = -3$:

$$4 = 0 + x_2 \qquad 0 = -3 + y_2$$
$$4 = x_2 \qquad 3 = y_2$$

The coordinates of B are **(4,3)**.

9. The equation of a circle with center at (h,k) and radius r is of this form:

$$(x - h)^2 + (y - k)^2 = r^2$$

For a center at $(1,3)$ and a radius of 2, $h = 1$, $k = 3$, and $r = 2$:

$$(x - 1)^2 + (y - 3)^2 = 2^2$$
$$= 4$$

The correct choice is **(2)**.

10. If a point is on the graph of $x + 3y = 13$, its coordinates must satisfy this equation.

Try each choice in turn by substituting the given coordinates in the equation $x + 3y = 13$:

(1) (4,4): $4 + 3(4) \stackrel{?}{=} 13$

                     $4 + 12 \stackrel{?}{=} 13$

                     $16 \neq 13$ $(4,4)$ is *not* on the graph.

(2) (–2,3): $-2 + 3(3) \stackrel{?}{=} 13$

                    $-2 + 9 \stackrel{?}{=} 13$

                     $7 \neq 13$ $(-2,3)$ is *not* on the graph.

(3) (–5,6): $-5 + 3(6) \stackrel{?}{=} 13$

                    $-5 + 18 \stackrel{?}{=} 13$

                   $13 = 13\checkmark$ $(-5,6)$ *is* on the graph.

(4) (4,–3): $4 + 3(-3) \stackrel{?}{=} 13$

                    $4 - 9 \stackrel{?}{=} 13$

                    $-5 \neq 13$ $(4,-3)$ is *not* on the graph.

The correct choice is **(3)**.

11. If an equation of a straight line is in the form $y = mx + b$, then m represents the slope.

To get the given equation, $4y = 3x + 16$, into the $y = mx + b$ form, divide each term by 4:

$$y = \frac{3}{4}x + 4$$

Here, $m = \dfrac{3}{4}$.

The slope is $\dfrac{3}{4}$.

12. If two lines are parallel, they must have the same slope.

If an equation of a line is in the form $y = mx + b$, then m represents the slope. The given equation, $y = 3x - 4$, is in the $y = mx + b$ form with $m = 3$, so its slope is 3.

Examine each choice in turn to see which one also has a slope of 3:

(1) $y = 4x - 3$ is in the $y = mx + b$ form with $m = 4$.
    Therefore, it is *not* parallel to $y = 3x - 4$.

(2) $y = 3x + 4$ is in the $y = mx + b$ form with $m = 3$.
    Therefore, it *is* parallel to $y = 3x - 4$.

(3) $y = -3x + 4$ is in the $y = mx + b$ form with $m = -3$.
    Therefore, it is *not* parallel to $y = 3x - 4$.

(4) $y = 3$ can be written in the $y = mx + b$ form as $y = 0x + 3$.
    In this form, $m = 0$. Therefore, $y = 3$ is *not* parallel to $y = 3x - 4$.

The correct choice is **(2)**.

13. An equation of a line whose slope is m and whose y-intercept is b can be written in the form $y = mx + b$.

In this case, $m = 2$ and $b = -2$. Hence, an equation of the line is $y = 2x = -2$.

The correct choice is **(4)**.

7. SYSTEMS OF EQUATIONS AND INEQUALITIES

Systems of Linear Equations

7.1 SOLVING A SYSTEM OF LINEAR EQUATIONS GRAPHICALLY: AN EXAMPLE

Solve graphically each of the following systems of equations:

$$y = x + 3$$
$$2x + y = 3$$

Draw the graphs of both equations on the same set of axes. The coordinates of the point of intersection of the two graphs represent the solution to the system.

Step 1. Draw the graph of $y = x + 3$. Set up a table of values by choosing three convenient values for x and substituting them in the equation to find the corresponding values of y:

x	$x + 3$	$= y$
0	$0 + 3$	$= 3$
3	$3 + 3$	$= 6$
-3	$-3 + 3$	$= 0$

Plot points (0,3), (3,6), and (–3,0). They should lie on a straight line. Draw this line; it is the graph of $y = x + 3$.

Step 2. Draw the graph of $2x + y = 3$. To do this, it is advisable to first rearrange the equation so that it is in a form in which it is solved for y in terms of x.

The given equation is: $2x + y = 3$

Add $-2x$ (the additive inverse of
$2x$) to both sides:

$$\underline{-2x \qquad = \quad -2x}$$
$$y = 3 - 2x$$

Set up a table of values by choosing three convenient values for x and substituting them in the equation to find the corresponding values of y:

x	$3 - 2x$		$= y$
0	$3 - 2(0)$	$= 3 - 0$	$= 3$
3	$3 - 2(3)$	$= 3 - 6$	$= -3$
-3	$3 - 2(-3)$	$= 3 + 6$	$= 9$

Plot points $(0,3)$, $(3,-3)$, and $(-3,9)$, as shown in the accompanying figure. They should lie in a straight line. Draw this line; it is the graph of $2x + y = 3$.

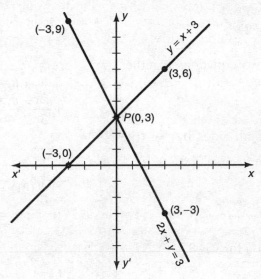

Step 3. Locate P, the point of intersection of the two graphs, which represents the common solution. The coordinates of P are $(0,3)$ or $x = 0$ and $y = 3$.

The solution to the system is $x = 0$, $y = 3$ or $\{(0,3)\}$.

CHECK: The solution is checked by substituting 0 for x and 3 for y in *both* of the two *original* equations to see whether they are satisfied:

$$
\begin{array}{ll}
y = x + 3 & 2x + y = 3 \\
3 \overset{?}{=} 0 + 3 & 2(0) + 3 \overset{?}{=} 3 \\
3 \overset{\checkmark}{=} 3 & 0 + 3 \overset{?}{=} 3 \\
& 3 \overset{\checkmark}{=} 3
\end{array}
$$

7.2 SOLVING A SYSTEM OF LINEAR EQUATIONS ALGEBRAICALLY: TWO EXAMPLES

Solve algebraically each of the following systems of equations that follow.

1. $x + y = 7$
$2x - y = 2$

Add the two equations together to eliminate y:

$$
\begin{array}{r}
x + y = 7 \\
2x - y = 2 \\
\hline
3x \quad\;\; = 9
\end{array}
$$

Multiply both sides by $\dfrac{1}{3}$ (the multiplicative inverse of 3):

$$
\frac{1}{3}(3x) = \frac{1}{3}(9)
$$
$$
x = 3
$$

The solution for x is **3**.

2. $2x + y = 6$
$x = 3y + 10$

Rearrange the second equation by adding $-3y$ (the additive inverse of $3y$) to both sides:

$$
\begin{array}{r}
x = 3y + 10 \\
-3y = -3y
\end{array}
$$

$$x - 3y = 10$$

Multiply each term in the first equation by 3:

$$6x + 3y = 18$$

Add the new form of the second equation, thus eliminating y:

$$x - 3y = 10$$

$$\overline{7x \qquad = 28}$$

Multiply both sides by $\frac{1}{7}$ (the multiplicative inverse of 7):

$$\frac{1}{7}(7x) = \frac{1}{7}(28)$$

Substitute 4 for x in the first equation:

$$x = 4$$
$$2(4) + y = 6$$
$$8 + y = 6$$

Add -8 (the additive inverse of 8) to both sides:

$$\underline{-8 \qquad = -8}$$
$$y = -2$$

The solution is $x = 4$, $y = -2$.

Systems of Linear Inequalities

7.3 SOLVING A SYSTEM OF LINEAR INEQUALITIES GRAPHICALLY: AN EXAMPLE

Solve graphically.

a. $\quad y > x + 4$
$\quad x + y \leq 2$

b. Determine which point lies in the solution set:
  (1) (2,3)   (3) (0,6)
  (2) (−5,2)   (4) (−1,0)

a. Step 1. Graph the solution set of the inequality $y > x + 4$. The graph of the inequality $y > x + 4$ is represented by all the points on the coordinate plane for which y, the ordinate, is greater than $x + 4$. Hence, first draw the graph of the line for which $y = x + 4$; having this graph will enable you to locate the region for which $y > x + 4$.

Select any three convenient values for x and substitute them in the equation $y = x + 4$ to find the corresponding values of y:

x	$x + 4$	$= y$
0	$0 + 4$	$= 4$
3	$3 + 4$	$= 7$
-4	$-4 + 4$	$= 0$

Plot points $(0,4)$, $(3,7)$, and $(-4,0)$. Draw a *dotted line* through these three points to get the graph of $y = x + 4$. The dotted line is used to signify that points on it are *not* part of the solution set of the inequality $y > x + 4$.

To find the *region* or *half-plane* on one side of the line $y = x + 4$ that represents $y > x + 4$, select a test point, say $(1,8)$, one one side of the line. Substituting in the inequality $y > x + 4$ gives $8 > 1 + 4$, or $8 > 5$, which is true. Thus, the side of the line on which $(1,8)$ lies (above and to the left) is the region representing $y > x + 4$. Shade it with cross-hatching extending up and to the left, as shown in the accompanying figure.

Step 2. Graph the solution set of $x + y \leq 2$. This graph is represented by all the points on the coordinate plane for which $x + y < 2$ in addition to the points on the line for which $x + y = 2$. Hence, the line $x + y = 2$ is first graphed. To make it convenient to find points on the line, solve for y in terms of x:

The given equation is: $\qquad\qquad\qquad\qquad x + y = 2$

Add $-x$ (the additive inverse of x) to both sides of the equation:

$$\underline{-x \qquad = \qquad -x}$$
$$y = 2 - x$$

Set up a table by selecting any three convenient values for x and substituting them in the equation $y = 2 - x$ to find the corresponding values of y:

x	$2 - x$	$= y$
0	$2 - 0$	$= 2$
3	$2 - 3$	$= -1$
5	$2 - 5$	$= 3$

Plot points $(0,2)$, $(3,-1)$, and $(5,-3)$. Draw a *solid line* through these three points to get the graph of $x + y = 2$. The solid line is used to signify that points on it are part of the solution set of $x + y \leq 2$.

To find the *region* or *half-plane* on one side of the line $x + y = 2$ for which $x + y < 2$, select a test point, say $(-2,-1)$, on one side of the line. Substituting $(-2,-1)$ in the inequality $x + y < 2$ results in $-2 - 1 < 2$, or $-3 < 2$, which is true. Thus, the side of the line where $(-2,-1)$ is located (below and to the left) is the region representing $x + y < 2$. Shade this region with cross-hatching extending down and to the left, as shown in the accompanying figure.

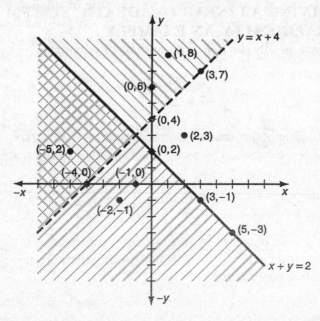

b. If a point is in the solution set, it will lie in the region in which the graph shows *both* sets of cross-hatching; points in this region satisfy *both* inequalities. Test each point in turn by locating it on the graph:

    (1) (2,3) does *not* lie in the cross-hatched area.

    (2) (−5,2) lies in the region with both sets of cross-hatching. Therefore, it is in the solution set of the graph.

    (3) (0,6) lies in a region cross-hatched only for $y > x + 4$; it satisfies this inequality but *not* the other.

    (4) (−1,0) lies in a region cross-hatched only for $x + y \leq 2$; it satisfies this inequality but *not* the other.

The only point in the solution set of the graph is **(−5,2)**.

Linear-Quadratic Systems

7.4 SOLVING A LINEAR-QUADRATIC SYSTEM GRAPHICALLY: AN EXAMPLE

Solve graphically the following systems of equations:

$$y = x^2 - 6x + 5$$
$$y + 7 = 2x$$

Step 1. Graph the second-degree equation. The graph of $y = x^2 - 6x + 5$ is a parabola. The x-coordinate of the turning point of the parabola is

$$x = -\frac{b}{2a} = -\frac{-6}{2(1)} = 3$$

Set up a table of values by picking three consecutive integer values of x on either side of $x = 3$ and substituting them in the equation to find their y-coordinates:

x	$x^2 - 6x\ +5$	$= y$
0	$0^2 - 6(0) + 5$	$= 5$
1	$1^2 - 6(1) + 5 = 1\ -\ 6 + 5$	$= 0$
2	$2^2 - 6(2) + 5 = 4\ -12 + 5$	$= -3$
3	$3^2 - 6(3) + 5 = 9\ -18 + 5$	$= -4$
4	$4^2 - 6(4) + 5 = 16 - 24 + 5$	$= -3$
5	$5^2 - 6(5) + 5 = 25 - 30 + 5$	$= 0$
6	$6^2 - 6(6) + 5 = 36 - 36 + 5$	$= 5$

Plot points (0,5), (1,0), (2,–3), (3,–4), (4,–3), (5,0), and (6,5) on a coordinate plane, and connect them with a smooth curve that has the shape of a parabola. The graph is labeled $y = x^2 - 6x + 5$ on the accompanying figure.

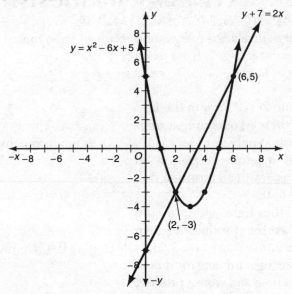

Step 2. Graph the first-degree equation using the same set of axes. The graph of $y + 7 = 2x$ is a line. Since $y = 2x - 7$, the line crosses the y-axis at $(0,-7)$. To find a second point on the line, pick any convenient value of x. For example, if $x = 5$, then $y = 2x - 7 = 2(5) - 7 = 3$. Plot $(0,-7)$ and $(5,3)$. Then connect the two points with a straight line. The graph is labeled $y + 7 = 2x$ on the accompanying figure.

Step 3. Find the coordinates of the points(s) at which the graphs intersect.

The solutions are **(2,–3)** and **(6,5)**.

7.5 SOLVING A LINEAR-QUADRATIC SYSTEM ALGEBRAICALLY: AN EXAMPLE

Solve algebraically the following system of equations:

$$y = x^2 - 6x + 5$$
$$y + 7 = 2x$$

Solve the first-degree equation for y by subtracting 7 from each side of the equation:

$$y + 7 = 2x$$
$$y = 2x - 7$$

Substitute $2x - 7$ for y in the first equation, thereby obtaining a quadratic equation only in x:

$$2x - 7 = x^2 - 6x + 5$$

Put the quadratic equation in standard form with all terms on one side equal to 0:

$$0 = x^2 - 8x + 12$$

Factor the right side of the equation as the product of two binomials:

$$0 = (x + ?)(x + ?)$$

Find the two missing numbers whose sum is –8 and whose product is +12. The two numbers are –2 and –6:

$$0 = (x - 2)(x - 6)$$

If the product of two factors is 0, either factors may equal 0:

$$x - 2 = 0 \quad \text{or} \quad x - 6 = 0$$
$$x = 2 \quad \text{or} \quad x = 6$$

To find the corresponding values of y, substitute each of the solutions for x in the original first-degree equation:

Let $x = 2$ Let $x = 6$

$$y + 7 = 2x \qquad y + 7 = 2x$$
$$= 2(2) \qquad\qquad = 2(6)$$
$$y = 4 - 7 \qquad y = 12 - 7$$
$$= -3 \qquad\qquad = 5$$

The solutions are **(2,–3)** and **(6,5)**.

CHECK: Substitute each pair of values of x and y in *both* of the *original* equations to verify that both equations are satisfied.

Let $x = 2$ and $y = -3$:

$$y = x^2 - 6x + 5 \qquad\qquad y + 7 = 2x$$
$$-3 \overset{?}{=} 2^2 - 6(2) + 5 \qquad -3 + 7 \overset{?}{=} 2(2)$$
$$-3 \overset{?}{=} 4 - 12 + 5 \qquad\qquad 4 \overset{\checkmark}{=} 4$$
$$-3 \overset{?}{=} -8 + 5$$
$$-3 \overset{\checkmark}{=} -3$$

Let $x = 6$ and $y = 5$:

$$5 \overset{?}{=} 6^2 - 6(6) + 5 \qquad 5 + 7 \overset{?}{=} 2(6)$$
$$5 \overset{?}{=} 36 - 36 + 5 \qquad\qquad 12 \overset{\checkmark}{=} 12$$
$$5 \overset{?}{=} 0 + 5$$
$$5 \overset{\checkmark}{=} 5$$

Practice Exercises

1. If the graphs of $x^2 + y^2 = 4$ and $y = -4$ are drawn on the same axes, what is the total number of points common to both graphs?
(1) 1 (2) 2 (3) 3 (4) 0

2. What is a solution for the system of equations $x - y = 2$ and $y = 2x - 4$?
(1) (0,2) (2) (2,0) (3) (3,2) (4) (4,2)

Solutions

1. Since the equation $x^2 + y^2 = r^2$ represents the graph of a circle with center at the origin and a radius of r, the graph of $x^2 + y^2 = 4$ (or $x^2 + y^2 = 2^2$) is a circle with center at the origin and a radius of 2.

The graph of the equation $y = -4$ is a straight line parallel to the x-axis and 4 units below it.

The two graphs do not intersect, so they have no points in common.

The correct choice is **(4)**.

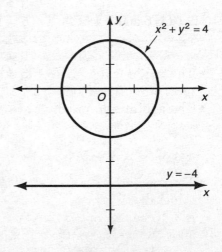

2. The given system of equations is:

$$x - y - 2$$
$$y = 2x - 4$$

Eliminate y in the first equation by substituting its equal, $2x - 4$:

$$x - (2x - 4) = 2$$

Remove the parentheses by taking the opposite of each term that is inside the parentheses:

$$x - 2x + 4 = 2$$

Solve for x:

$$-x + 4 = 2$$
$$x = 2$$

Find the corresponding value of y by substituting 2 for x in the second equation:

$$y = 2(2) - 4 = 0$$

The solution is $(2,0)$.

The correct choice is **(2)**.

8. PROBABILITY AND STATISTICS

Statistics

8.1 KEY STATISTICS

Each of the following statistics is a single number that helps describe how the individual data values in a list are distributed.

- The **mode** is the number in a list of data values that occurs the most often. The mode of $\{2, 2, 3, 3, 3, 4\}$ is 3.
- The **arithmetic mean** or **average** of a set of data values is the sum of the data values divided by the number of values. The average of $\{1, 2, 3, 4, 5\}$ is $\dfrac{1+2+3+4+5}{3} = \dfrac{15}{3} = 5$.
- The **median** is the number in the middle position of a list of data values that is arranged in size order. The median of $\{1, 2, 3, 4, 5\}$ is 3. If the list contains an even number of data values, the median is the average of the two middle values. The median of $\{2, 5, 8, 14, 15, 19\}$ is $\dfrac{8+14}{2} = 11$.

- The **range** is the difference obtained by subtracting the lowest value from the highest value in the list.
- The **pth percentile** is the data point at or below which $p\%$ of the data values fall. The median is the 50th percentile.

8.2 QUARTILES

Quartiles divide a set of data values into four equal groups.
- The **middle quartile** is the median.
- The **lower quartile** is the median of the values that fall below the overall median.
- The **upper quartile** is the median of the values that fall above the overall median.

8.3 BOX-AND-WHISKER PLOT

A **box-and-whisker plot** visually summarizes a set of values using five key values: the lowest value (L), the lower quartile (Q_1), the median (Q_2), the upper quartile (Q_3), and the highest value (H), as shown in the accompanying diagram.

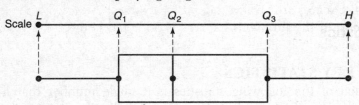

8.4 FREQUENCY TABLES AND HISTOGRAMS: AN EXAMPLE

The table below gives the distribution of test scores for a class of 20 students.

Test Score Interval	Number of Students (frequency)
91–100	1
81–90	3
71–80	3
61–70	7
51–60	6

a. Draw a *frequency histogram* for the given data.
b. Which interval contains the median?
c. Which interval contains the lower quartile?

a.

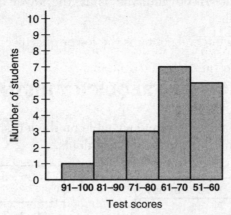

b. Find the total number of students by adding all the frequencies: $1 + 3 + 3 + 7 + 6 = 20$.

If the scores are arranged in order of size, the *median* is the middle score if the number of scores is odd, and is the score midway between the two middle scores if, as here, there is an even number of scores. Since there are 20 scores, the median score will lie midway between the 10th and 11th scores.

Counting up from the bottom, there are six scores in the 51–60 interval. To count up to the 10th score, four more are needed from the next (the 61–70) interval; five more are needed from the same interval to reach the 11th score. Both the 10th and 11th scores are in the 61–70 interval. The score midway between them (the median) must also lie in this interval.

The **61–70** interval contains the median.

c. The *lower quartile* is the score separating the lowest one-quarter of all scores from the remaining three-quarters.

Since there are 20 scores in all and $\frac{1}{4}(20) = 5$, the lowest five scores comprise the lowest one-quarter; the remaining 15 scores comprise the upper three-quarters. The lower quartile is the score midway between the 5th and 6th scores.

Counting up from the bottom, the 5th and 6th scores both lie in the lowest (the 51–60) interval. Thus, the lower quartile is in the 51–60 interval.

The **51–60** interval contains the lower quartile.

8.5 CUMULATIVE FREQUENCY HISTOGRAMS: AN EXAMPLE

The accompanying frequency table on the left shows the distribution of weight, in pounds, of 32 students.

Interval	Frequency
169–179	9
140–159	8
120–139	6
100–119	2
80–99	7

Interval	Cumulative Frequency
80–179	
80–159	
80–139	
80–119	
80–99	

a. On another sheet of paper, copy and complete the accompanying cumulative frequency table on the right, using the data given in the frequency table.

b. Construct a cumulative frequency histogram using the table completed in part *a*.

a.

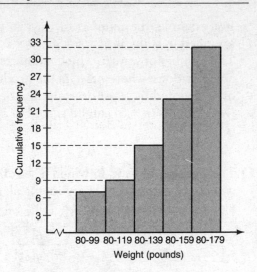

Interval	Cumulative Frequency
80–179	32
80–159	23
80–139	15
80–119	9
80–99	7

b. See the accompanying histogram.

Counting Methods

8.6 MULTIPLICATION PRINCIPLE OF COUNTING

The **multiplication principle of counting** states that, if an event can happen in n ways and a second event can happen next in m ways, the first event followed by the second event can happen in $n \times m$ ways. For example, if John has five shirts, four ties, and six pairs of pants, then the number of different outfits, each consisting of a shirt, a tie, and a pair of pants, that John can put together is $5 \times 4 \times 6 = 120$.

8.7 COUNTING ORDERED ARRANGEMENTS OF OBJECTS

• The number of ways in which n objects can be arranged in a line is $n!$ (n factorial), where $n!$ represents the product of consecutive whole numbers from n down to 1:

$$n! = n \times (n-1) \times (n-2) \times \cdots \times 1$$

For example, the number of ways in which five students can be arranged in a line is $5! = 5 \times 4 \times 3 \times 2 \times 1 = 120$.

- The number of ways in which n objects can be arranged in r available positions, where $r \leq n$, is the product of the r greatest factors of n, denoted by $_nP_r$. For example, the number of ways in which five students ($n = 5$) can be seated in a row of three chairs ($r = 3$) is

$$_5P_3 = 5 \times 4 \times 3 = 60$$

- The number of ways in which a set of n objects can be arranged when x objects in the set are identical, y objects are identical, and z objects are identical is

$$\frac{n!}{x! \cdot y! \cdot z!}$$

8.8 COUNTING UNORDERED ARRANGEMENTS OF OBJECTS

In a **combination** of objects, unlike a permutation, the identity rather than the order of the objects matters. For example, if three people will be selected from a group of five people to serve on a committee, the names of the three people, rather than their order, is important.

- From a group of five people, the number of different committees of three people that can be formed is represented by $_5C_3$, which is read as "the combination of five people taken three at a time."
- With a scientific calculator, $_5C_3$ evaluates to 10. You can also evaluate $_5C_3$ by using the formula

$$_nC_r = \frac{n!}{r!(n-r)!}$$

where, in this case, $n = 5$ and $r = 3$.

- The multiplication principle of counting applies also to combinations. Suppose there are six pens and seven books on a desk. If Raymond wants to select four pens and three books from the desk, the number of selections that are possible is $_6C_4 \times _7C_3 = 525$.

8.9 TREE DIAGRAMS

The set of all possible outcomes for one event, or a series of events, is called a **sample space**. The multiplication principle of counting tells how many outcomes are in a sample space, but does not describe the individual outcomes. When you need to know the identity of each of the outcomes in a sample space, you can draw a **tree diagram** or make an organized list.

Suppose a jar contains a red (R) marble, a blue (B) marble, and a white (W) marble. After a marble is selected at random and its color is noted, it is put back into the jar. Then another marble is picked at random and its color is noted. Using the multiple principle of counting, you know that 3 × 3 or 9 different outcomes are possible. These outcomes can be described as shown in the accompanying tree diagram:

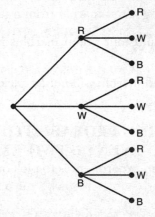

The set of all possible outcomes can also be described by making a list of ordered pairs of the form (color of first marble, color of second marble):

(R, R)	(W, R)	(B, R)
(R, W)	(W, W)	(B, W)
(R, B)	(W, B)	(B, B)

Probability

8.10 MEANING OF PROBABILITY

A **probability** value expresses the likelihood that a future event will happen. If an event is certain to happen, its probability is 1. If an event is impossible, its probability is 0. All other events have probability values between 0 and 1.

- If an event E can happen in r of n equally likely ways, the probability that it will happen is defined as follows:

$$P(E) = \frac{\text{number of favorable outcomes}}{\text{total number of possible outcomes}} = \frac{r}{n}.$$

- If the probability that an event will happen is p, the probability that the event will *not* happen is $1 - p$. For example, the probability of picking a red marble from a jar that contains only three red marbles and four green marbles is $\frac{3}{3+4}$ or $\frac{3}{7}$. The probability that a red marble will *not* be picked is $1 - \frac{3}{7} = \frac{4}{7}$.

8.11 DETERMINING PROBABILITIES OF COMPOUND EVENTS: TWO EXAMPLES

1. The first step of an experiment is to pick one number from the set {1,2,3}. The second step is to pick one number from the set {1,4,9}.

a. Draw a tree diagram or list the sample space of all possible pairs of outcomes.

b. Determine the probability that:
  - (1) both numbers are the same
  - (2) the second number is the square of the first
  - (3) both numbers are odd

a. The tree diagram is shown first. The tree diagram contains three branches leading from "Start" since either 1, 2, or 3 may be picked on the first step. For each of the "first-step" branches, three "second-step" branches lead to 1, 4, or 9 as the possible second number to be picked:

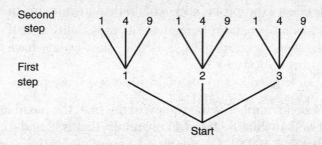

The accompanying table representing the sample space contains two columns, one for each of the two steps in the experiment. The column for the first step shows the possible numbers, 1, 2, and 3, for that step. In the second column, each number in the first column is paired with 1, 4, and 9, the possible numbers that may be picked in the second step.

First Step	Second Step
1	1
1	4
1	9
2	1
2	4
2	9
3	1
3	4
3	9

b. Probability of an event $= \dfrac{\text{number of successful outcomes}}{\text{total number of possible outcomes}}$.

(1) The only way in which both numbers can be the same is if 1 is chosen both times. The tree diagram shows that only one path

leads to two selections of 1—the leftmost path. There are nine possible paths in all. If the sample-space table is used, there is only one line in which both columns contain 1—the first line. There are nine lines in all. Therefore, the probability that both numbers are the same = $\dfrac{1}{9}$.

The probability is $\dfrac{1}{9}$.

(2) If the second number is the square of the first, the possible selections are 1 and 1^2 (that is, 1 and 1), 2 and 2^2 (that is, 2 and 4), and 3 and 3^2 (that is, 3 and 9). On the tree diagram, one path leads to 1 and 1, a second path leads to 2 and 4, and a third path leads to 3 and 9. Thus, three paths of the possible total of nine are favorable cases. If the sample space is used to obtain the information, the combinations of 1 and 1, 2 and 4, and 3 and 9 are shown on one line each, a total of three lines (favorable cases) of the nine possible lines. Therefore, the probability that the second number is the square of the first = $\dfrac{3}{9}$.

The probability is $\dfrac{3}{9}$.

(3) The tree diagram shows that the paths leading to both odd numbers are $1-1$, $1-9$, $3-1$, and $3-9$. Thus, four possible paths of the total of nine represent the successful outcomes. In the table for the sample space, the lines containing 1 and 1, 1 and 9, 3 and 1, and 3 and 9 represent the successful outcomes of lines containing both odd numbers. There are nine lines in all. Therefore, the probability that both numbers are odd is $\dfrac{4}{9}$.

The probability is $\dfrac{4}{9}$.

2. The probability of the Bears beating the Eagles is $\frac{1}{2}$. The probability of the Bears beating the Rams is $\frac{3}{5}$. What is the probability that the Bears will win both games?

The probability of $\frac{1}{2}$ means that the Bears will beat the Eagles in $\frac{1}{2}$ of the games they play against the Eagles. Similarly, the probability of $\frac{3}{5}$ means that the Bears will beat the Rams in $\frac{3}{5}$ of the games they play against the Rams. If a Bears-Eagles game is followed by a Bears-Rams game, then the $\frac{1}{2}$ of the times the Bears win the first game will be followed $\frac{3}{5}$ of the time by their winning the second game.

Thus, the probability that the Bears will win both games is $\frac{1}{2} \times \frac{3}{5} = \frac{3}{10}$.

The probability is $\frac{3}{10}$.

8.12 DETERMINING PROBABILITIES USING COMBINATIONS: AN EXAMPLE

A committee of six is to be chosen from four juniors and five seniors.

a. What is the probability that the committee will include the same number of juniors and seniors?

b. What is the probability that the committee will include all five seniors?

c. What is the probability that the committee will include no senior?

Probability of an event occurring

$$= \frac{\text{number of favorable outcomes}}{\text{total possible number of outcomes}}$$

a. If a committee of six is to include the same number of juniors and seniors, then it must include three of each.

The number of possible combinations of three juniors chosen from the four available juniors is:

$$_4C_3 = \frac{4(3)(2)}{3(2)(1)} = \frac{4(\cancel{3})(\cancel{2})}{\cancel{3}(\cancel{2})(1)} = \frac{4}{1} = 4$$

The number of possible combinations of three seniors chosen from the five available seniors is:

$$_5C_3 = \frac{5(4)(3)}{3(2)(1)} = \frac{5(\cancel{4})(\cancel{3})}{\cancel{3}(\cancel{2})(1)} = \frac{10}{1} = 10$$

Each of the four combinations of three juniors may be combined with each of the 10 combinations of three seniors to form 4×10, or 40, successful outcomes for a committee that includes the same number of juniors as seniors. The total possible number of outcomes for a committee of six is the number of combinations of all nine students taken six at a time:

$$_9C_6 = \frac{9(8)(7)(6)(5)(4)}{6(5)(4)(3)(2)(1)} = \frac{\cancel{9}(\cancel{8})(7)(\cancel{6})(\cancel{5})(\cancel{4})}{\cancel{6}(\cancel{5})(\cancel{4})(\cancel{3})(\cancel{2})(1)} = \frac{84}{1} = 84$$

The probability of a committee that will include the same number of juniors and seniors $\frac{40}{84}$ or $\frac{10}{21}$.

The probability is $\frac{\mathbf{10}}{\mathbf{21}}$.

b. There is only one combination of all five seniors. The committee of six may be formed by combining this one combination of all five seniors with any one of the four juniors. Thus, there are four successful outcomes for a committee of six that will include all five seniors. The total number of outcomes for the selection of such a committee is 84 (see part ***a***).

The probability of a committee of six containing all five seniors

$= \dfrac{4}{84}$ or $\dfrac{1}{21}$.

The probability is $\dfrac{1}{21}$.

c. To include no senior, all six members of the committee would have to be juniors, but there are only four juniors from which to choose. Thus, there is no possible favorable outcome for a committee with no senior.

The probability of a committee of six with no senior $= \dfrac{0}{84}$ or 0.

The probability is **0**.

Practice Exercises

1. If a boy has five shirts and three pairs of pants, how many possible outfits, each consisting of one shirt and one pair of pants, can be chosen?

2. How many different five-person committees can be selected from nine people?

3. Which expression is *not* equivalent to $_8C_5$?
  (1) 56   (3) $_8C_3$

  (2) $_8C_5$ (4) $\dfrac{8 \cdot 7 \cdot 6}{3 \cdot 2 \cdot 1}$

4. How many different arrangements of seven letters can be made using the letters in the name "ULYSSES"?

5. How many different five-letter permutations can be formed from the letters of the word "DITTO"?

  (1) 5!   (3) $\dfrac{5!}{2!}$

  (2) $(5-2)!$ (4) $_5P_2$

6. A candy jar has four red (R) gumdrops, five green (G) gumdrops, and one black (B) gumdrop. Without looking, Kim reaches into the jar and chooses one gumdrop. Then, without replacing this gumdrop, Kim chooses a second gumdrop. The accompanying tree diagram represents all possible outcomes with the probability value on each branch:

a. Find the values of x, y, and z.

b. Find the probability that
(1) both gumdrops chosen are green
(2) one of the gumdrops chosen is red and the other is green

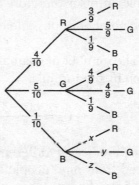

7. A committee of five is to be chosen from six freshmen and eight sophomores. What is the probability that the committee will include two freshmen and three sophomores?

8. The accompanying cumulative frequency histogram shows the scores that 24 students received on an English test. How many students had scores between 71 and 80?

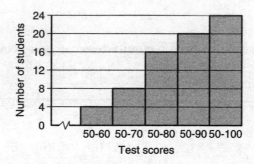

Solutions

1. The boy has five choices for shirts. For each of these five choices, he has three choices for pairs of pants. The total number of possible outfits is therefore $5 \times 3 = 15$.

The number of possible outfits is **15**.

2. The number of combinations of n things taken r at a time, $_nC_r$, is given by the formula:

$$_nC_r = \frac{n(n-1)(n-2)(n-3)\ldots\text{to } r \text{ factors}}{r!}$$

The number of different five-person committees that can be selected from nine people is the number of combinations of nine things taken five at a time.

Let $n = 9$ and $r = 5$:

$$_9C_5 = \frac{9(8)(7)(6)(5)}{5(4)(3)(2)(1)}$$

$$= \frac{9(\overset{2}{8})(7)(\overset{\overset{1}{2}}{6})(\overset{1}{5})}{\overset{}{5}(\overset{}{4})(\overset{}{3})(\overset{}{2})(1)}$$

$$= \frac{126}{1} = 126$$

126 different five-person committees can be selected.

3. By definition:

$$_nC_r = \frac{n(n-1)(n-2)(n-3)\ldots\text{to } r \text{ factors}}{r!}$$

Evaluate $_8C_5$:

$$_8C_5 = \frac{8(7)(6)}{3(2)(1)} = \frac{8(7)(\overset{\overset{1}{2}}{6})}{\overset{}{3}(\overset{}{2})(1)} = \frac{56}{1} = 56$$

Choice (1), 56, and choice (4), $\dfrac{8 \cdot 7 \cdot 6}{3 \cdot 2 \cdot 1}$, are both equivalent to $_8C_5$. $_nC_r$ is always equivalent to $_nC_{n-r}$. If $n = 8$ and $r = 5$, then $n - r = 3$. Thus, $_8C_5$ is equivalent to $_8C_3$, which is choice (3).

By definition, $_nP_r = n(n-1)(n-2)(n-3)$ ... to r factors. Therefore, $_8P_5 = 8(7)(6)(5)(4)$. The first three factors alone are equivalent to the numerator of $_8C_5$. Therefore, $_8P_5$, which is choice (2), is *not* equivalent to $_8C_5$; in fact, it is much larger.

The correct choice (2).

4. The number of different arrangements of p things where r are alike is $\dfrac{p!}{r!}$.

The name "ULYSSES" contains seven letters but there are three S's.

For $p = 7$ and $r = 3$:

$$\frac{p!}{r!} = \frac{7!}{3!} = \frac{7(6)(5)(4)(3)(2)(1)}{3(2)(1)}$$

$$= \frac{7(6)(5)(4)\cancel{(3)}\cancel{(2)}\cancel{(1)}}{\cancel{3(2)(1)}}$$

$$= \frac{840}{1} = 840$$

840 different arrangements are possible.

5. The number of permutations, P, of n things taken n at a time when r are alike is given by the formula

$$P = \frac{n!}{r!}$$

The word "DITTO" has five letters but there are two T's. For the number of permutations of the letters in "DITTO," let $n = 5$ and $r = 2$; then

$$P = \frac{5!}{2!}$$

The correct choice is **(3)**.

6.

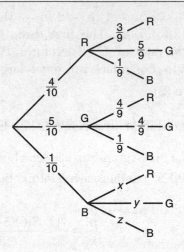

a. On the accompanying tree diagram each of branches x, y, and z corresponds to the probability that a gumdrop of a particular color will be selected on the second pick, given that a black gumdrop was selected, without replacement, on the first pick.

In each case, the replacement set for the second pick consists of nine gumdrops: four red (R), five green (G), and zero black (B). Hence:

$$x = P(\text{R on second pick}) = \frac{\text{number of red gumdrops}}{\text{total number of gumdrops}} = \frac{4}{9}$$

$$y = P(\text{G on second pick}) = \frac{\text{number of green gumdrops}}{\text{total number of gumdrops}} = \frac{5}{9}$$

$$z = P(\text{B on second pick}) = \frac{\text{number of black gumdrops}}{\text{total number of gumdrops}} = \frac{4}{9} = 0$$

Thus, $x = \dfrac{4}{9}, y = \dfrac{5}{9}, z = 0$.

b. (1) Multiplying the probability values encountered in the tree diagram as a path is traced along the branch, from the center to the first G and then to the second G, gives the probability of selecting a green gumdrop followed by another green gumdrop:

$$P(\text{G on both picks}) = \frac{5}{10} \times \frac{4}{9} = \frac{20}{90}$$

The probability that both gumdrops selected are green is $\frac{20}{90}$.

(2) If a red and a green gumdrop are selected, either color may be selected first. Thus:

$$P(\text{R and G}) = P(\text{R followed by G}) + P(\text{G followed by R})$$

Multiplying the probability values encountered in the tree diagram as a path is traced along the branch, from the center to R and then to G, gives the probability of selecting a red gumdrop followed by a green gumdrop:

$$P(\text{R followed by G}) = \frac{4}{9} \times \frac{5}{10} = \frac{20}{90}$$

Similarly, multiplying the probability values along the branch, from the center to G and then R, gives the probability of selecting a green gumdrop followed by a red gumdrop:

$$P(\text{G followed by R}) = \frac{5}{10} \times \frac{4}{9} = \frac{20}{90}$$

Thus:

$$P(\text{R and G}) = \frac{20}{90} + \frac{20}{90} = \frac{40}{90}$$

The probability that a red and a green gumdrop will be selected is $\frac{40}{90}$.

7. It is given that a committee of five is to be chosen from a group of 14 students (6 freshmen and 8 sophomores). Without considering order, r objects can be selected from a group of n objects in $_nC_r$ ways, where $_nC_r$ can be evaluated using a scientific calculator or the formula $_nC_r = \dfrac{n!}{r!(n-r)!}$.

To find the total number of ways any five people can be chosen from a total of 14, $n = 14$ and $r = 5$. Use your calculator to determine that $_{14}C_5 = 2002$.

To find the number of ways to select two freshmen from the six freshmen in the group, let $n = 6$ and $r = 2$, and determine that $_6C_2 = 15$.

Similarly, to find the number of ways to select three sophomores from the eight sophomores in the group, let $n = 8$, and $r = 3$, and determine that $_8C_3 = 56$.

The probability of selecting two freshmen and three sophomores is the product of the number of ways to select the two groups divided by 2002, the total number of five-person committees possible, calculated above.

$$P(2 \text{ freshmen and 3 sophomores}) = \frac{_6C_2 \cdot \, _8C_3}{_{14}C_5} = \frac{15 \cdot 56}{2002} = \frac{840}{2002}$$

The probability of including two freshmen and three sophomores is $\dfrac{840}{2002}$.

8. The cumulative frequency histogram shows that eight students received scores between 50 and 70 and 16 students received scores between 50 and 80. Hence, $16 - 8$ or 8 students received scores between 71 and 80.

8 students received scores between 71 and 80.

Glossary of Terms

abscissa The x-coordinate of a point in the coordinate plane. The abscissa of the point $(2, 3)$ is 2.

absolute value The absolute value of a number x, denoted by $|x|$, is its distance from zero on the number line. Thus, the absolute value of a nonzero number is the number or its opposite—whichever is nonnegative.

acute angle An angle whose degree measure is less than 90.

acute triangle A triangle that contains three acute angles.

additive inverse The opposite of a number. The additive inverse of a number x is $-x$ since $x + (-x) = 0$.

adjacent angles Two angles that have the same vertex, share a common side, but do not have any interior points in common.

alternate interior angles Two interior angles that lie on opposite sides of a transversal.

altitude A segment that is perpendicular to the side to which it is drawn.

angle The union of two rays that have the same endpoint.

antecedent The part of a conditional statement that follows the word "if." Sometimes the term "hypothesis" is used in place of "antecedent."

associative property The mathematical law that states that the order in which three numbers are added or multiplied does not matter.

average See *Mean*.

axes The two number lines that intersect at right angles to form a coordinate plane.

base angles of an isosceles triangle The two congruent angles that include the base of an isosceles triangle.

base of a power The number that is repeated as a factor in a product. In $3^4 = 3 \cdot 3 \cdot 3 \cdot 3 = 81$, 3 is the base and 4 is the exponent since 3 appears as a factor 4 times.

biconditional A statement of the form p if and only if q, denoted by $p \leftrightarrow q$, which is true only when p and q have the same truth value.

binomial A polynomial with two unlike terms.

bisector of an angle A ray that divides the angle into two angles that have the same degree measure.

bisector of a segment A line, ray, or segment that contains the midpoint of the given segment.

coefficient The number that multiplies the literal factors of a monomial. The coefficient of $-5x^2y$ is -5.

collinear points Points that lie on the same line.

commutative property The mathematical law that states that the order in which two numbers are added or multiplied does not matter.

complementary angles Two angles whose degree measures add up to 90.

compound statement A statement in logic that is formed by combining two or more simple statements using logical connectives.

conditional statement A statement that has the form "If p then q," denoted by $p \rightarrow q$. It is always true except in the case in which p is true and q is false.

congruent angles Angles that have the same degree measure.

congruent figures Figures that have the same size and the same shape. The symbol for congruence is $\cong$.

congruent polygons Two polygons having the same number of sides are congruent if their vertices can be paired so that all corresponding sides have the same length and all corresponding angles have the same degree measure.

congruent triangles Two triangles are congruent if any one of the following conditions is true: (1) the sides of one triangle are congruent to the corresponding sides of the other triangle ($SSS \cong SSS$); (2) two sides and the included angle of one triangle are congruent to the corresponding parts of the other triangle ($SAS \cong SAS$); (3) two angles and the included side of one triangle are congruent to the corresponding parts of the other triangle ($ASA \cong ASA$); (4) two angles and the side opposite one of these angles of one triangle are congruent to the corresponding parts of the other triangle ($AAS \cong AAS$).

congruent segments Line segments that have the same length.

conjugate pair The sum and difference of the same two terms, as in $a + b$ and $a - b$.

conjunct Each of the individual statements that comprise a conjunction.

conjunction A statement of the form p and q, denoted by $p \wedge q$, which is true only when p and q are true at the same time.

consequent The part of a conditional statement that follows the word "then." Sometimes the word "conclusion" is used in place of "consequent."

constant A quantity that is fixed in value. In the equation $y = x + 3$, x and y are variables and 3 is a constant.

contradiction A compound statement that is always false.

contrapositive The conditional statement formed by negating and then interchanging both parts of a conditional statement. The contrapositive of $p \rightarrow q$ is $\sim q \rightarrow \sim p$.

converse The conditional statement formed by interchanging both parts of a conditional statement. The converse of $p \rightarrow q$ is $q \rightarrow p$.

coordinate The real number that corresponds to the position of a point on a number line.

coordinate plane The region formed by a horizontal number line and vertical number line intersecting at their zero points.

corresponding angles Pairs of angles that lie on the same side of a transversal, one of which is an interior angle and the other is an exterior angle. In congruent or similar polygons, corresponding angles are pairs of angles formed at matching vertices.

corresponding sides In congruent or similar polygons, corresponding sides are pairs of sides whose endpoints are matching vertices.

cumulative frequency The sum of all frequencies from a given data point up to and including another data point.

cumulative frequency histogram A histogram whose bar heights represent the cumulative frequency at stated intervals.

data Items of information.

degree A unit of angle measure that is defined as 1/360th of one complete rotation of a ray about its vertex.

degree of a monomial The sum of the exponents of its variable factors. For example, the degree of $3x^4$ is 4 and the degree of $-4xy^2$ is 3 since 1 (the power of x) plus 2 (the power of y) equals 3.

degree of a polynomial The greatest degree of its monomial terms. For example, the degree of $x^2 - 4x + 5$ is 2.

dilation A transformation in which a figure is enlarged or reduced in size based on a center and a scale factor.

direct variation A set of ordered pairs in which the ratio of the second member to the first member of each ordered pair is the same nonzero number. Thus, if y varies directly as x then $\frac{y}{x} = k$ or, equivalently, $y = kx$, where k is a nonzero number called the *constant of variation*.

disjunct Each of the statements that comprise a disjunction.

disjunction A statement of the form p or q, denoted by $p \vee q$, which is true when p is true, q is true, or both p and q are true.

distributive property of multiplication over addition For any real numbers a, b, and c, $a(b + c) = ab + ac$ and $(b + c)a = ba + ca$.

domain The set of all possible replacements for a variable.

equation A statement that two quantities have the same value.

equilateral triangle A triangle whose three sides have the same length.

equivalent equations Two equations that have the same solution set. Thus, $2x = 6$ and $x = 3$ are equivalent equations.

event A particular subset of outcomes from the set of all possible outcomes of a probability experiment. In flipping a coin, one event is getting a head; another event is getting a tail.

exponent A number that indicates how many times another number, called the base, is used as a factor. In 3^4, the number 4 is the exponent and tells the number of times the base 3 is used as a factor in a product. Thus, $3^4 = 3 \cdot 3 \cdot 3 \cdot 3 = 81$.

extremes In the proportion $\frac{a}{b} = \frac{c}{d}$, the terms a and d are the *extremes*.

factor A number or variable that is being multiplied in a product.

factoring The process by which a number or polynomial is written as the product of two or more terms.

factoring completely Factoring a number or polynomial into its prime factors.

factorial n Denoted by $n!$ and defined for any positive integer n as the product of consecutive integers from n to 1. Thus, $5! = 5 \cdot 4 \cdot 3 \cdot 2 \cdot 1 = 120$.

FOIL The rule for multiplying two binomials horizontally by forming the sum of the products of the first terms (F), the outer terms (O), the inner terms (I), and the last terms (L) of each binomial.

formula An equation that shows how one quantity depends on one or more other quantities.

fraction A number that has the form $\frac{a}{b}$, where a is a real number called the numerator and b is a real number called the *denominator*, provided $b \neq 0$.

frequency The number of times a data value appears in a list.

Fundamental Counting Principle If event A can occur in m ways and event B can occur in n ways, then both events can occur in m times n ways.

graph of an equation The set of all points on a number line or in the coordinate plane that are solutions of the equation.

Greatest Common Factor (GCF) The GCF of two or more monomials is the monomial with the greatest coefficient and the variable factors of the greatest degree that are common to all the given monomials. The GCF of $8a^2b$ and $20ab^2$ is $4ab$.

histogram A vertical bar graph whose bars are adjacent to each other.

hypotenuse The side of a right triangle that is opposite the right angle.

identity An equation that is true for all possible replacements of the variable. The equation $2(x + 1) = 2x + 2$ is an identity.

image In a geometric transformation, the point or figure that corresponds to the original point or figure.

inequality A sentence that compares two quantities using an inequality relation such as < (is less than), ≤ (is less than or equal to), > (is greater than), ≥ (is greater than or equal to), or ≠ (is unequal to).

integer A number from the set $\{\ldots -3, -2, -1, 0, 1, 2, 3, \ldots\}$.

inverse The statement formed by negating both the antecedent and consequent of a conditional statement. Thus, the inverse of $p \rightarrow q$ is $\sim p \rightarrow \sim q$.

irrational number A number that cannot be expressed as the quotient of two integers.

isosceles triangle A triangle in which at least two sides have the same length.

leg of a right triangle Either side of a right triangle that is not opposite the right angle.

like terms Terms that differ only in their numerical coefficients. For example, $2x$ and $3x$ are like terms.

line Although an undefined term in geometry, it can be described as a continuous set of points that describes a straight path that extends indefinitely in two opposite directions.

linear equation An equation in which the greatest exponent of a variable is 1. A linear equation can be put into the form $Ax + By = C$, where A, B, and C are constants and A and B are not both zero.

line reflection A transformation in which each point P that is not on line ℓ is paired with a point P' on the opposite side of line ℓ so that line ℓ is the perpendicular bisector of $\overline{PP'}$. If P is on line ℓ, then P is paired with itself.

line segment Part of a line that consists of two different points on the line, called *endpoints*, and all points on the line that are between them.

line symmetry A figure has line symmetry when a line ℓ divides the figure into two parts such that each part is the reflection of the other part in line ℓ.

logical connectives The conjunction ($\wedge$), disjunction ($\vee$), conditional ($\rightarrow$), and biconditional ($\leftrightarrow$) of two statements.

mean The mean or average of a set of n data values is the sum of the data values divided by n.

means In the proportion $\dfrac{a}{b} = \dfrac{c}{d}$, the terms b and c are the *means*.

median The middle value when a set of numbers is arranged in size order. If the set has an even number of values, then the median is the average of the middle two values. For example, the median of 2, 4, 8, 11, and 24 is 8. The median of 7, 11, 23, and 29 is $\dfrac{11 + 23}{2} = 17$.

mode The data value that occurs most frequently in a given set of data.

monomial A number, variable, or their product.

multiplicative inverse The reciprocal of a nonzero number.

negation The negation of statement p is the statement, denoted by $\sim p$, that has the opposite truth value of p.

obtuse angle An angle whose degree measure is greater than 90 and less than 180.

obtuse triangle A triangle that contains an obtuse angle.

open sentence A sentence whose truth value cannot be determined until its placeholders are replaced with values from the replacement set.

opposite rays Two rays that have the same endpoint and form a line.

ordered pair Two numbers that are written in a definite order.

ordinate The y-coordinate of a point in the coordinate plane. The ordinate of the point $(2, 3)$ is 3.

origin The zero point on a number line.

outcome A possible result in a probability experiment.

parallel lines Lines in the same plane that do not intersect. When a third line, called a *transversal*, intersects two parallel lines, every pair of angles formed are either congruent or supplementary.

parallelogram A quadrilateral that has two pairs of parallel sides. In a parallelogram, opposite sides are congruent, opposite angles are congruent, consecutive angles are supplementary, and each diagonal divides the parallelogram into two congruent triangles.

perfect square A rational number whose square root is also rational. The perfect square integers from 1 to 100 are 1, 4, 9, 25, 36, 49, 64, 81, and 100. The number $\frac{4}{9}$ is an example of a fraction that is a perfect square since $\sqrt{\frac{4}{9}} = \frac{2}{3}$ and $\frac{2}{3}$ is rational.

permutation An ordered arrangement of objects. For example, AB and BA represent two different permutations of the letters A and B.

perpendicular lines Lines that intersect at right angles.

plane Although undefined in geometry, it can be described as a flat surface that extends indefinitely in all directions.

point Although undefined in geometry, it can be described as indicating location with no size.

polygon A simply closed curve whose sides are line segments.

polynomial A monomial or the sum or difference of two or more monomials.

postulate A statement whose truth is accepted without proof.

power A product of identical factors written in the form x^n where the number n, called the *exponent*, gives the number of times the common factor x, called the *base*, is used in the product.

prime factorization The factorization of a polynomial into factors each of which are divisible only by itself and 1 (or -1).

probability of an event The number of ways in which the event can occur divided by the total number of possible outcomes.

proportion An equation that states that two ratios are equal. In the proportion $\frac{a}{b} = \frac{c}{d}$, the product of the means equals the product of the extremes. Thus, $b \cdot c = a \cdot d$.

Pythagorean Theorem The square of the length of the hypotenuse of a right triangle is equal to the sum of the squares of the lengths of the legs of the right triangle.

quadrant One of four rectangular regions into which the coordinate plane is divided.

quadratic equation An equation that can be put into the form $ax^2 + bx + c = 0$, provided $a \neq 0$.

quadratic polynomial A polynomial like $x^2 - 3x + 4$ whose degree is 2.

quadrilateral A polygon with four sides.

radical (square root) sign The symbol $\sqrt{}$ that denotes the positive square root of a nonnegative number.

radicand The expression that appears underneath a radical sign.

ratio A comparison of two numbers by division. The ratio of a to b is the fraction $\frac{a}{b}$, provided $b \neq 0$.

rational number A number that can be written in the form $\frac{a}{b}$ where a and b are integers and $b \neq 0$. Decimals in which a set of digits endlessly repeat, like $.25000 \ldots \left(= \frac{1}{4} \right)$ and $.33333 \ldots \left(= \frac{1}{3} \right)$ represent rational numbers.

ray The part of a line that consists of a point, called an *endpoint*, and the set of points on one side of the endpoint.

real number A number that is a member of a set that consists of all rational and irrational numbers.

reciprocal The reciprocal of a nonzero number x is $\frac{1}{x}$. For example, the reciprocal of 3 is $\frac{1}{3}$ and the reciprocal of $-\frac{2}{5}$ is $-\frac{5}{2}$. The product of a nonzero number and its reciprocal is always 1.

rectangle A parallelogram with four right angles.

replacement set The set of values that a variable may have.

rhombus A parallelogram with four sides that have the same length.

right angle An angle whose degree measure is 90.

right triangle A triangle that contains a right angle.

root A number that makes an equation a true statement.

rotation A transformation in which a point or figure is moved about a fixed point a given number of degrees.

rotational symmetry A figure has rotational symmetry if it can be rotated so that its image coincides with the original figure.

scalene triangle A triangle in which no two sides have the same length.

scientific notation A number is in scientific notation when it is expressed as the product of a number between 1 and 10 and a power of 10. The number 81,000 written in scientific notation is 8.1×10^4.

similar polygons Two polygons with the same number of sides are similar if their vertices can be paired so that corresponding angles have the same measure and the lengths or corresponding sides are in proportion.

similar triangles If two triangles have two pairs of corresponding angles that have the same degree measure, then the triangles are similar.

slope A measure of the steepness of a nonvertical line. The slope of a horizontal line is 0, and the slope of a vertical line is undefined.

slope formula The slope of a nonvertical line that contains the points (x_1, y_1) and (x_2, y_2) is given by the formula $\frac{y_2 - y_1}{x_2 - x_1}$.

slope-intercept form An equation of a line that has the form $y = mx + b$ where m is the slope of the line and b is the y-intercept.

solution Any value from the replacement set of a variable that makes an open sentence true.

solution set The collection of all values from the replacement set of a variable that makes an open sentence true.

square A rectangle whose four sides have the same length.

square root The square root of a nonnegative number n is one of two identical numbers whose product is n. Thus, $\sqrt{9} = 3$ because $3 \times 3 = 9$.

statement Any sentence whose truth value can be assessed as true or false, but not both.

success Any favorable outcome of a probability experiment.

supplementary angles Two angles are supplementary if the sum of their degree measures is 180.

system of equations Two or more equations whose solution is the set of values that makes each equation true at the same time. For the system of equations $x + y = 5$ and $y = x + 1$, the solution is $(2, 3)$ since $x = 2$ and $y = 3$ makes both equations true.

tautology A compound statement that is true regardless of the truth values of its component statements.

theorem A generalization in mathematics that can be proved.

transformation The process of "moving" each point of a figure according to some given rule.

translation A transformation in which each point of a figure is moved the same distance and in the same direction.

transversal A line that intersects two other lines in two different points.

trapezoid A quadrilateral with exactly one pair of parallel sides.

tree diagram A diagram whose branches describe the different possible outcomes in a probability experiment.

triangle A polygon with three sides.

trinomial A polynomial with three unlike terms.

truth value Every statement has a truth value of either true *or* false, but not both.

undefined term A term that can be described but not defined. The terms *point*, *line*, and *plane* are undefined in geometry.

unlike terms Terms that do not have the same variable factors.

variable The symbol, usually a letter, that represents an unspecified member of a given set called the *replacement set*.

vertical angles A pair of nonadjacent angles formed by two intersecting lines.

x-axis The horizontal axis in the coordinate plane.
x-coordinate The first number in an ordered pair.

y-axis The vertical axis in the coordinate plane.
y-coordinate The second number in an ordered pair.
y-intercept The y-coordinate of the point at which the graph of an equation intersects the y-axis.

Regents Examinations, Answers, and Self-Analysis Charts

Math A
Sample Test

In June 1999, the New York State Board of Regents began offering a new math test, known as Math A Regents Examination. Eventually, two math tests, A and B, will replace the current I, II, and III exams. The last administration of Sequential Math Course I will be in January 2002, Sequential Course II is scheduled to end with the January 2003 test, and the final Sequential Course III will be in January 2004. The Math B Exam was first given in June 2001.

Many schools will continue to offer the Course I, II, and III Regents Examinations until they are phased out, but many other schools began offering the Math A Exam in June 1999.

To help prepare students for the Math A Exam, the following sample exam (with answer key) was developed by the Board of Regents. The sample test is followed in this book by actual Math A exams, starting with the June 1999 test.

The following information may be helpful in preparing for the new exams.

- Students must answer all questions.
- Formulas will not be provided.
- Students should have access to a straightedge and a compass.
- Scientific calculators must be available to all students.

PART I

Answer all questions in this part. Each correct answer will receive two (2) credits. No partial credit will be allowed. Write your answers in the spaces provided. [40]

1 For what value of x will 8 and x have the same mean (average) as 27 and 5?

(1) 1.5   (3) 24
(2) 8   (4) 40   1 _____

2 If $12x = 4(x + 5)$, then x equals

(1) $\dfrac{1}{12}$ (3) 1.25

(2) $\dfrac{5}{8}$ (4) 2.5   2 _____

3 The image of point (3,4) when reflected in the y-axis is

(1) $(-3,-4)$ (3) $(3,-4)$
(2) $(-3,4)$ (4) $(4,3)$ 3 _____

4 If $n - 3$ is an even integer, what is the next larger consecutive even integer?

(1) $n - 5$ (3) $n + 1$
(2) $n - 1$ (4) $n + 2$ 4 _____

5 If $2a^2 - 6a + 5$ is subtracted from $3a^2 - 2a + 3$, the result is

(1) $5a^2 - 8a + 8$ (3) $-a^2 - 4a + 2$
(2) $a^2 + 4a - 2$ (4) $a^2 - 8a + 8$ 5 _____

6 Which is a factor of $x^2 + 5x - 24$?

   (1) $(x + 4)$ (3) $(x + 3)$

   (2) $(x - 4)$ (4) $(x - 3)$ 6 _____

7 When $6y^6 - 18y^3 - 12y^2$ is divided by $-3y^2$, the quotient is

   (1) $2y^4 - 6y^2 - 4y$ (3) $-2y^4 + 6y + 4$

   (2) $3y^4 + 6y + 4$ (4) $-2y^3 - 6y^2 - 4y$ 7 _____

8 If 0.0154 is expressed in the form 1.54×10^n, n is equal to

   (1) -2 (3) 3

   (2) 2   (4) -3 8 _____

9 How many integer values of x are there so that x, 5, and 8 could be the lengths of the sides of a triangle?

   (1) 6   (3) 3

   (2) 9   (4) 13   9 _____

10 In the diagram below, m$\angle BCD = 130$ and m$\angle B = 20$. What is m$\angle A$?

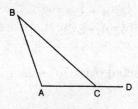

   (1) 50   (3) 110

   (2) 70   (4) 150   10 _____

11 What is the distance between points $A(7,3)$ and $B(5,-1)$?

(1) $\sqrt{10}$ (3) $\sqrt{14}$

(2) $\sqrt{12}$ (4) $\sqrt{20}$ 11 ____

12 "If Mary and Tom are classmates, then they go to the same school." Which statement below is logically equivalent?

(1) If Mary and Tom do not go to the same school, then they are not classmates.
(2) If Mary and Tom are not classmates, then they do not go to the same school.
(3) If Mary and Tom go to the same school, then they are classmates.
(4) Mary and Tom go to the same school, then they are not classmates.   12 ____

13 For what value of t is $\dfrac{1}{\sqrt{t}} < \sqrt{t} < t$ true?

(1) 1   (3) -1

(2) 0   (4) 4   13 ____

14 There are 12 tomato plants in a garden. Each plant has 7 branches and each branch has four (4) tomatoes growing on it. If one-third of the tomatoes are picked, how many tomatoes were picked?

(1) 23   (3) 224

(2) 112   (4) 336   14 ____

15 In the diagram below, $\overline{AB}$ is parallel to $\overline{CD}$. Transversal $\overline{EF}$ intersects $\overline{AB}$ and $\overline{CD}$ at G and H, respectively. If $m\angle AGH = 4x$ and $m\angle GHD = 3x + 40$, what is the value of x?

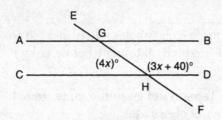

(1) 20   (3) 80

(2) 40   (4) 160   15 _____

16 Laura goes shopping. She spends one-fourth of her money on a pair of shorts, and one-third of her remaining money for a belt. If Laura has $42 left after these two purchases, how much money did she have when she started shopping?

(1) $84   (3) $144

(2) $126   (4) $504   16 _____

17 The distance between points P and Q is eight (8) units. How many points are equidistant from P and Q and also three (3) units from P?

(1) 1   (3) 0

(2) 2   (4) 4   17 _____

18 The accompanying histogram shows the scores of students on a Math A test.

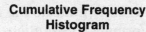

Cumulative Frequency Histogram

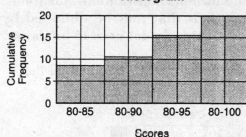

How many students have scores of 96 to 100?

(1) 55   (3) 5

(2) 20   (4) 4   18 _____

19 The expression $\sqrt{150}$ is equivalent to

(1) $25\sqrt{6}$ (3) $5\sqrt{6}$

(2) $15\sqrt{10}$ (4) $6\sqrt{5}$ 19 _____

20 Erica cannot remember the correct order of the four digits in her ID number. She does remember that the ID number contains the digits 1, 2, 5, and 9. What is the probability that the first three digits of Erica's ID number will all be odd numbers?

(1) $\dfrac{1}{4}$ (3) $\dfrac{1}{2}$

(2) $\dfrac{1}{3}$ (4) $\dfrac{3}{4}$ 20 _____

PART II

Answer all questions in this part. Each correct answer will receive two (2) credits. Clearly indicate the necessary steps, including appropriate formula substitutions, diagrams, graphs, charts, etc. Calculations that may be obtained by mental arithmetic or the calculator do not need to be shown. [10]

21 The graph below shows the hair colors of all the students in a class.

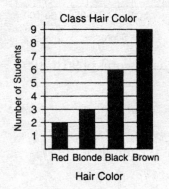

Class Hair Color

What is the probability that a student chosen at random from this class has black hair?   21 _____

22 In the figure shown below, each dot is one unit from an adjacent horizontal or vertical dot.

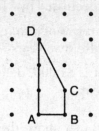

Find the number of square units in the area of quadrilateral *ABCD*. Show how you arrived at your answer.

22 _____

23 A design was constructed by using two rectangles *ABDC* and *A'B'D'C'*. Rectangle *A'B'D'C'* is the result of a translation of rectangle *ABDC*. The table of translations is shown below. Find the coordinates of points *B* and *D'*.

Rectangle *ABDC*	Rectangle *A'B'D'C'*
A (2,4)	A' (3,1)
B	B' (−5,1)
C (2,−1)	C' (3,−4)
D (−6,−1)	D'

23 B _____

D' _____

24 Mr. Cash bought d dollars worth of stock. During
the first year, the value of the stock tripled. The next
year, the value of the stock decreased by $1200.

(a) Write an expression in terms of d to represent
the value of the stock after two years.   24 a) ___

(b) If an initial investment is $1,000, determine its
value at the end of 2 years.   24 b) ___

25 The tailgate of a truck is 2 feet above the ground.
The incline of a ramp used for loading the truck is
11°, as shown below.

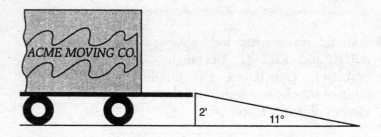

Find, to the *nearest tenth of a foot*, the length of
the ramp.   25 _____

PART III

**Answer all questions in this part. Each correct answer will
receive three (3) credits. Clearly indicate the necessary
steps, including appropriate substitutions, diagrams, graphs,
charts, etc. Calculations that may be obtained by mental
arithmetic or the calculator do not need to be shown.** [15]

26 On his first 5 biology tests, Bob received the following scores: 72, 86, 92, 63, and 77. What test score
must Bob earn on his sixth test so that his average
(mean score) for all six tests will be 80? Show how
you arrived at your answer.   26 _____

27 The figure below represents the distances traveled by car A and car B in 6 hours.

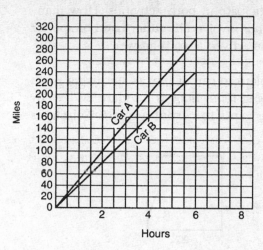

Which car is going faster and by how much? Explain how you arrived at your answer.

27 _____

28 A total of 800 votes were cast in an election. The table below represents the votes that were received by the candidates. Candidate D got at least 30 votes more than Candidate E. What is the *least* number of votes that Candidate D could have received? Show how you arrived at your answer.

Candidate	Number of votes
A	213
B	328
C	39
D	x
E	y

28 _____

29 In a school of 320 students, 85 students are in the band, 200 students are on sports teams, and 60 students participate in both activities. How many students are *not* involved in either band or sports? Show how you arrived at your answer.

29 ____

30 Ms. Brown plans to carpet part of her living room floor. The living room floor is a square 20 feet by 20 feet. She wants to carpet a quarter-circle as shown below.

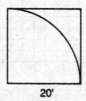

20'

Find, to the *nearest square foot*, what part of the floor will remain uncarpeted. Show how you arrived at your answer.

30 ____

PART IV

Answer all questions in the part. Each correct answer will receive four (4) credits. Clearly indicate the necessary steps, including appropriate substitutions, diagrams, graphs, charts, etc. Calculations that may be obtained by mental arithmetic or the calculator do not need to be shown. [20]

31 Two video rental clubs offer two different rental fee plans:
Club *A* charges $12 for membership and $2 for each rented video.
Club *B* has a $4 membership fee and charges $4 for each rented video.

The graph below represents the total cost of renting videos from Club *A*.

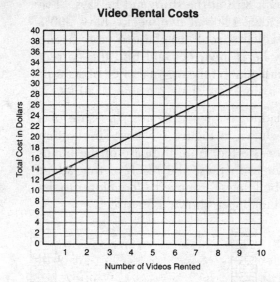

Video Rental Costs

(a) On the same set of *xy*-axes, draw a line to represent the total cost of renting videos from Club *B*.
(b) For what number of video rentals is it less expensive to belong to Club *A*? Explain how you arrived at your answer.

31 b) _____

32 Jed bought a generator that will run for 2 hours on a liter of gas. The gas tank on the generator is a rectangular prism with dimensions 20 cm by 15 cm by 10 cm as shown below.

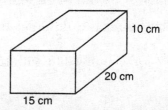

10 cm
20 cm
15 cm

If Jed fills the tank with gas, how long will the generator run? Show how you arrived at your answer.

32 _____

33 A clothing store offers a 50% discount at the end
of each week that an item remains unsold. Patrick
wants to buy a shirt at the store and he says, "I've
got a great idea! I'll wait two weeks, have 100%
off, and get it for free!" Explain to your friend
Patrick why he is incorrect and find the correct
percent of discount on the original price of a shirt.   33 ____

34 A 10-foot ladder is placed against the side of a
building as shown in Figure 1 below. The bottom
of the ladder is 8 feet from the base of the building. In order to increase the reach of the ladder
against the building, it is moved 4 feet closer to the
base of the building as shown in Figure 2.

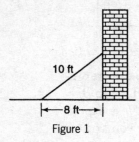

Figure 1

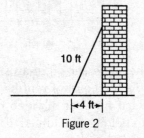

Figure 2

To the *nearest foot*, how much further up the
building does the ladder now reach? Show how
you arrived at your answer.   34 ____

35 A corner is cut off a 5″ by 5″ square piece of paper. The cut is x inches from a corner as shown below.

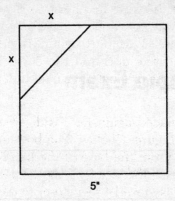

(a) Write an equation, in terms of x, that represents the area, A, of the paper after the corner is removed.

35 a) _____

(b) What value of x will result in an area that is $\dfrac{7}{8}$ of the area of the original square piece of paper? Show how you arrived at your answer.

35 b) _____

Answers

Math A Sample Exam

Answer Key

PART I

1. (3)	**8.** (1)	**15.** (2)
2. (4)	**9.** (2)	**16.** (1)
3. (2)	**10.** (3)	**17.** (3)
4. (2)	**11.** (4)	**18.** (4)
5. (2)	**12.** (1)	**19.** (3)
6. (4)	**13.** (4)	**20.** (1)
7. (3)	**14.** (2)	

PART II

21. $\frac{6}{20}$

22. 2

23. B (–6,4)
D (–5,–4)

24. **a.** $3d - 1200$
b. \$1800

25. 10.5

PART III

26. 90

27. Car A by 10 mph

28. 125

29. 95

30. 86

PART IV

31. **a.** slope of 4, y-intercept of 4
b. 5 or more

32. 6 hours

33. 75%, with explanation

34. 3 feet

35. **a.** $A = 25 - \dfrac{x^2}{2}$ or equivalent

36. **b.** 2.5

Answers Explained

PART I

1. The mean or average of two numbers is the sum of the two numbers divided by 2.

Since it is given that 8 and x have the same mean (average) as 27 and 5:

$$\frac{8+x}{2} = \frac{27+5}{2} = \frac{32}{2}$$

When two fractions with the same denominator are equal, the numerators of the two fractions must also be equal:

$$8+x = 32$$

Isolate x by subtracting 8 from each side of the equation:

$$x = 32 - 8 = 24$$

The correct choice is (**3**).

2. The given equation is:

$$12x = 4(x+5)$$

Remove the parentheses by multiplying each term inside the parentheses by the number in front of the parentheses:

$$12x = 4x + 20$$

Subtract $4x$ from each side of the equation:

$$12x - 4x = 20$$
$$8x = 20$$

Divide each side of the equation by 8:

$$x = \frac{20}{8} = 2.5$$

The correct choice is (**4**).

3. The image of a point that is reflected in the y-axis is on the opposite side of the y-axis as the original point and the same distance from the y-axis. In general, the image of $P(x,y)$ after a reflection in the y-axis is $P'(-x,y)$.

When reflected in the y-axis, the image of $(3,4)$ is $(-3,4)$, as shown in the accompanying diagram.

The correct choice is (**2**).

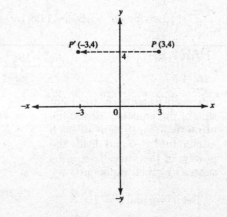

4. In a list of consecutive even integers, each even integer after the first is 2 more than the number that comes before it, as in

$$\ldots, -4, -2, 0, 2, 4, 6, \ldots.$$

Thus, if $n - 3$ represents an even integer, the next larger consecutive even integer is $(n - 3) + 2$ or $n - 1$.

The correct choice is (**2**).

5. Write the polynomial to be subtracted, $2a^2 - 6a + 5$, underneath the other polynomial, $3a^2 - 2a + 3$, aligning like terms in the same column:

$$3a^2 - 2a + 3$$
$$-$$
$$\underline{2a^2 - 6a + 5}$$

Convert to an addition example by changing the sign of each term of the polynomial to be subtracted to its opposite:

$$3a^2 - 2a + 3$$
$$+$$
$$\underline{-2a^2 + 6a - 5}$$

Add like terms in each column:

$$a^2 + 4a - 2$$

The correct choice is (**2**).

6. The given quadratic trinomial, $x^2 + 5x - 24$, can be factored as the product of two binomials that has the form $(x + a)(x + b)$. The letters a and b stand for the two integers whose sum is +5, the coefficient of the x-term in $x^2 + 5x - 24$, and whose product is -24, the constant term in $x^2 + 5x - 24$.

Since $(+8) + (-3) = +5$ and $(+8)(-3) = -24$, the two factors of $x^2 + 5x - 24$, are $(x + 8)$ and $(x - 3)$. The set of answer choices includes $(x - 3)$, but not $(x + 8)$.

The correct choice is (**4**).

7. To divide $6y^6 - 18y^3 - 12y^2$ by $-3y^2$, divide each term of the polynomial by $-3y^2$:

$$\frac{6y^6 - 18y^3 - 12y^2}{-3y^2} = \frac{6y^6}{-3y^2} + \frac{-18y^3}{-3y^2} + \frac{-12y^2}{-3y^2}$$

For each fraction on the right side of the equation, divide the numerical coefficient of each numerator by -3 and divide the powers of the same base, y, by subtracting their exponents:

$$= -2y^{6-2} + 6y^{3-2} + 4y^{2-2}$$
$$= -2y^4 + 6y^1 + 4y^0$$

Let $y^1 = y$ and $y^0 = 1$:

$$= -2y^4 + 6y + 4$$

The correct choice is (**3**).

8. If 0.0154 is expressed in the form 1.54×10^n, then

$$1.54 \times 10^n = 0.0154.$$

Since 1.54 must be multiplied by $0.01 = \dfrac{1}{100} = \dfrac{1}{10^2} = 10^{-2}$ to obtain 0.0154, $n = -2$.

The correct choice is (**1**).

9. In any triangle, the length of each side is less than the sum of the lengths of the other two sides and greater than their difference. Thus, $x < 5 + 8$ and $x > 8 - 5$, where x is an integer.

Since 4, 5, 6, 7, 8, 9, 10, 11, and 12 satisfy the condition that $x < 13$ *and* $x > 3$, there are 9 possible integer values for x.

The correct choice is (**2**).

10. In a triangle, the measure of an exterior angle is equal to the sum of the measures of the two nonadjacent interior angles. Thus:

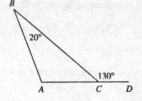

$$m\angle A + m\angle B = m\angle BCD$$

Substitute the given values for $\angle B$ and $\angle C$, shown in the accompanying diagram:

$$m\angle A + 20 = 130$$
$$m\angle A = 130 - 20$$
$$= 110$$

The correct choice is (**3**).

11. The distance, d, between points $A(x_A, y_A)$ and $B(x_B, y_B)$ is given by the formula

$$d = \sqrt{(x_B - x_A)^2 + (y_B - y_A)^2}.$$

To find the distance between points $A(7,3)$ and $B(5,-1)$, let $A(x_A, y_A) = (7,3)$ and $B(x_B, y_B) = (5,-1)$. Thus:

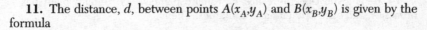

$$d = \sqrt{(x_B - x_A)^2 + (y_B - y_A)^2}$$
$$= \sqrt{(5 - 7)^2 = (-1 - 3)^2}$$
$$= \sqrt{(-2)^2 = (-4)^2}$$
$$= \sqrt{4 + 16}$$
$$= \sqrt{20}$$

The correct choice is (**4**).

12. A conditional statement $(p \rightarrow q)$ and its contrapositive $(\sim q \rightarrow \sim p)$ are logically equivalent since they always have the same truth value. To form the contrapositive of the conditional "If Mary and Tom are classmates, then they go to the same school," interchange and then negate the two parts of the conditional:

"If Mary and Tom do not go to the same school, then they are not classmates."

The correct choice is (**1**).

13. To find the answer choice that contains a value of t that makes $\dfrac{1}{\sqrt{t}} < \sqrt{t} < t$ a true statement, replace t in the inequality with each answer choice until you find the one that works.

Choice (1): Let $t = 1$. Since $\sqrt{1} = 1$, $\dfrac{1}{\sqrt{t}} < \sqrt{t} < t$ becomes $\dfrac{1}{1} < 1 < 1$, which is not true.

Choice (2): Let $t = 0$. Since $\dfrac{1}{\sqrt{0}}$ is not defined, $\dfrac{1}{0} < 0 < 0$ is not true.

Choice (3): Let $t = -1$. Since $\sqrt{-1}$ is not a real number, $\dfrac{1}{\sqrt{-1}} < \sqrt{-1} < -1$ is not true.

Choice (4): Let $t = 4$. Since $\sqrt{4} = 2$, $\dfrac{1}{2} < 2 < 4$, which is a true statement.

The correct choice is (**4**).

14. It is given that there are 12 tomato plants, each tomato plant has 7 branches, and each branch has 4 tomatoes. Then:

• Each tomato plant has 7×4 or 28 tomatoes.

• Since the garden has 12 tomato plants, it has 12×28 tomatoes.

• If one-third of the tomatoes are picked, the number of tomatoes picked is

$$\dfrac{1}{\cancel{3}} \times \overset{4}{\cancel{12}} \times 28 = 4 \times 28 = 112$$

The correct choice is (**2**).

15. It is given that, in the accompanying diagram, $\overline{AB}$ is parallel to $\overline{CD}$, m$\angle AGH = 4x$, and m$\angle GHD = 3x + 40$. Since alternate interior angles formed by parallel lines are equal in measure:

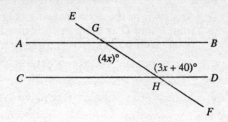

$$\text{m}\angle AGH = \text{m}\angle GHD$$
$$4x = 3x + 40$$
$$4x - 3x = 40$$
$$x = 40$$

The correct choice is (**2**).

16. Let m represent the amount of money Laura had when she started shopping. Then:

- After she spends one-fourth of her money on a pair of shorts, the amount of money remaining is $\frac{3}{4}m$.

- If Laura now spends one-third of the money she has left of a belt, the amount of money left after the two purchases is $\frac{2}{3} \times \frac{3}{4}m = \frac{6}{12}m = \frac{1}{2}m$.

- Since Laura has \$42 left after the two purchases, $\frac{1}{2}m = \$42$ so $m = \$42 \times 2 = \$84$.

Laura had \$84 when she started shopping.

The correct choice is (**1**).

17. It is given that the distance between points P and Q is 8 units. You need to find the number of points that satisfy the following two given conditions at the same time:

Condition 1: Points are equidistant from P and Q.

Condition 2: Points are 3 units from P.

To find the number of points that satisfy two conditions at the same time, sketch the locus of points that satisfies each condition on the same diagram. Then count the number of points at which the two loci intersect.

- As shown in the accompanying diagram, the locus of points equidistant from P and Q is the perpendicular bisector of the segment whose endpoints are P and Q.

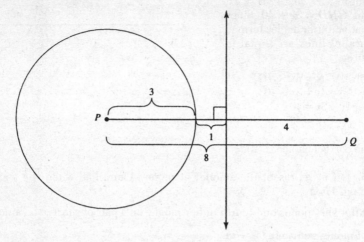

- The locus of points 3 units from P is a circle with the center at P and a radius of 3 units.

Since the circle does not intersect the perpendicular bisector of $\overline{PQ}$, there is 0 point that is equidistant from P and Q and also 3 units from P.

The correct choice is (**3**).

18. The accompanying histogram shows the scores of students on a Math A test. The height of each rectangular bar gives the total number of students with scores that fall in the interval at the base of that rectangle.

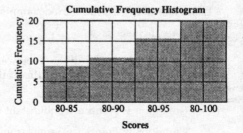

Since 16 students have scores in the interval 80–95 and 20 students have scores in the interval 80–100, 20 minus 16 or 4 students must have scores in the interval 96–100.

The correct choice is (**4**).

19. The given square root radical is: $\sqrt{150}$

Factor the radicand into two whole numbers so that one of the numbers is the greatest perfect square factor of 150: $\sqrt{25 \times 6}$

Write the radical over each factor: $\sqrt{25} \times \sqrt{6}$

Evaluate the perfect square radical: $5\sqrt{6}$

The correct choice is (**3**).

20. It is given that Erica's four-digit ID number contains the digits 1, 2, 5, and 9. The digits 1, 5, and 9 are odd. The probability that the first three digits of Erica's ID will all be odd numbers is the fraction obtained by dividing the number of ID numbers that can be formed in which the first three digits are all odd numbers by the total number of different ways in which the four digits can be arranged.

- The first step is to determine the number of different ways in which the available odd digits, 1, 5, and 9, can fill the first three positions of the ID number. There are three different odd digits that can appear in the first position of the ID number. Since each digit can be used only once, two odd digits remain that can appear in the second position of the ID number, leaving one odd digit for the third position. Thus only one possible digit, 2, is left to fill the last position of the ID number. Using the counting principle, there are $3 \times 2 \times 1 \times 1$ or 6 different four-digit ID numbers that can be formed from the digits 1, 2, 5, and 9 in which the first three digits are all odd numbers.

- If there is no restriction on where the odd digits can appear within the four-digit number, there are $4 \times 3 \times 2 \times 1$ or 24 different ways in which the four digits can be arranged.

Thus, the probability that the first three digits of Erica's ID number will all be odd numbers is $\dfrac{6}{24}$ or $\dfrac{1}{4}$.

The correct choice is (**1**).

PART II

21. The accompanying histogram shows the hair colors of all the students in a class. The base of each rectangular bar of the histogram gives the hair color, and the height of the bar tells the number of students who have that hair color. Thus:

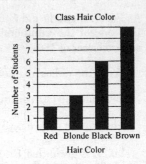

Class Hair Color

Hair Color	Number of Students
Red	2
Blonde	3
Black	6
Brown	9

Total = 20 students in the class

Since 6 of the 20 students have black hair, the probability that a student chosen at random from this class has black hair is $\dfrac{6}{20}$.

22. It is given that, in the accompanying figure, each dot is one unit from an adjacent horizontal or vertical dot.

To find the number of square units in the area of quadrilateral $ABCD$, divide the quadrilateral into a right triangle and a square by connecting C with the dot—call it E—that is one unit to its left. Then:

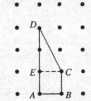

• Area of square $ABCD = AB \times BC = 1 \times 1 = 1$

• Area of right triangle CED

$$= \frac{1}{2} \times EC \times ED = \frac{1}{2} \times 1 \times 2 = 1$$

• Area of quadrilateral $ABCD$
  = Area of square $ABCD$ + Area of right triangle CED
  = 1   + 1
  = 2

The number of square units in the area of quadrilateral $ABCD$ is **2**.

23. In general, translation $T_{h,k}$ "slides" a point h units in the horizontal direction and k units in the vertical direction. Thus, after translation T, the image of (a,b) is $(a + h, b + k)$.

The accompanying table shows the results of translating rectangle $ABCD$ and obtaining rectangle $A'B'C'D'$ as its image.

Rectangle $ABDC$	Rectangle $A'B'C'D'$
$A(2,4)$	$A'(3,1)$
B	$B'(-5,1)$
$C(2,-1)$	$C'(3,-4)$
$D(-6,-1)$	D'

• First, find the translation rule. After translation $T_{h,k}$, the image of $A(2,4)$ is $A'(2 + h, 4 + k)$. Since it is given in the table that the image of $A(2,4)$ is $A'(3,1)$:

$$A'(2+h, 4+k) = A'(3,1)$$

Then

$$2 + h = 3 \quad \text{and} \quad 4 + k = 1$$
$$h = 1 \quad \text{and} \quad k = 1 - 4 = -3$$

Hence, the translation rule is $T_{1,-3}$.

• Let (x,y) represent the coordinates of B. After translation $T_{1,-3}$, the image of $B(x,y)$ is $B'(x + 1, y - 3)$. Hence:

$$B'(x+1, y-3) = B'(-5,1)$$

so

$$x + 1 = -5 \quad \text{and} \quad y - 3 = 1$$
$$x = -6 \quad \text{and} \quad y = 4$$

The coordinates of B are $(-6,4)$.

• After translation $T_{1,-3}$, the image of $D(-6,-1)$ is $D'(-6+1, -1-3) = D'(-5,-4)$.

Hence, the coordinates of D' are $(-5,-4)$.

24. It is given that Mr. Cash bought d dollars worth of stock that tripled in value during the first year. The next year the value of the stock decreased by $1200.

(a) The value of the stock after the first year was $3d$, and after 2 years was $3d - 1200$.

(b) If the initial investment was $1000, then letting $d = 1000$ gives

$$3d - 1200 = 3(1000) - 1200$$
$$= 3000 - 1200$$
$$= 1800$$

The value of an initial investment of $1000 at the end of 2 years is $**1800**.

25. It is given that, in the accompanying diagram, the tailgate of a truck is 2 feet above the ground and the incline of a loading ramp is 11°.

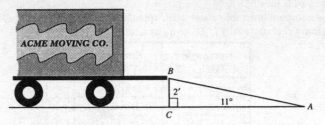

To find the length of the ramp to the nearest tenth of a foot, use the sine ratio in right triangle ACB:

$$\sin A = \frac{\text{leg opposite } \angle A}{\text{hypotenuse}}$$

$$\sin 11° = \frac{2}{AB}$$

Use a calculator to evaluate $\sin 11°$:

$$0.1908 = \frac{2}{AB}$$

$$AB \times 0.1908 = 2$$

$$AB = 2 \div 0.1908 = 10.482$$

The length of the ramp, to the *nearest tenth of a foot*, is **10.5** feet.

PART III

26. It is given that Bob received the scores 72, 86, 92, 63, and 77 on his first 5 biology tests. Let x represent the score Bob must receive on his next biology test so that his average (mean score) for all six tests will be 80.

Since the average of six scores is the sum of the six scores divided by 6:

$$\frac{x + 72 + 86 + 92 + 63 + 77}{6} = 80$$

$$\frac{x + 390}{6} = 80$$

$$x + 390 = 6 \cdot 80$$

$$= 480$$

$$x = 480 - 390$$

$$= 90$$

Bob must earn a score of **90** on his sixth test.

27. It is given that, in the accompanying figure, the lines drawn represent the distances traveled by car A and car B in 6 hours. To determine which car is going faster and by how much, find the average rate of speed of each car by dividing the change in distance along the vertical axis by the change in time along the horizontal axis, thereby obtaining the slope of each line.

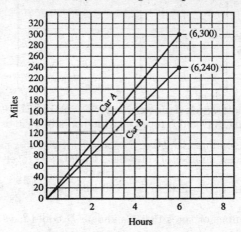

- From 0 to 6 hours, car A travels 300 miles, so its average rate of speed is

$$\frac{\text{change in distance}}{\text{change in time}} = \frac{300 - 0 \text{ miles}}{6 - 0 \text{ hours}} = 50\,\text{mph}$$

- From 0 to 6 hours, car B travels 240 miles, so its average rate of speed is

$$\frac{\text{change in distance}}{\text{change in time}} = \frac{240 - 0 \text{ miles}}{6 - 0 \text{ hours}} = 40\,\text{mph}$$

Car A is faster by $50 - 40 = \mathbf{10\,mph}$ than car B.

28. It is given that five candidates received a total of 800 votes, and the votes received by the candidates are shown in the accompanying table.

Candidate	Number of Votes
A	213
B	328
C	39
D	x
E	y

- Because candidates A, B, and C received a total of $213 + 328 + 39$ or 580 votes, candidates D and E received a total of $800 - 580$ or 220 votes.

- Since it is given that Candidate D got at least 30 votes more than Candidate E, the least number of votes that Candidate D received is $y + 30$.

- Hence, $y + (y + 30) = 220$, so $2y = 220 - 30$ and $y = \dfrac{190}{2} = 95$.

The *least* number of votes that Candidate D could have received is $y + 30 = 95 + 30 = \mathbf{125}$.

29. It is given that, in a school of 320 students, 85 students are in the band, 200 students are on sports teams, and 60 students participate in both activities. The number of students who are *not* involved in either band or sports can be obtained by subtracting the total number of students involved in just one of these activities from 320.

Method I: Use a Venn Diagram.

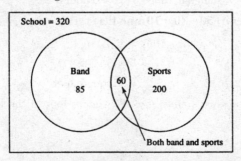

The accompanying Venn diagram shows that $85 + 200 - 60 = 225$ different students are involved in just one of the two activities. In other words, 225 students belong to the band or belong to sports teams, but do not belong to both. Hence, of the 320 students who attend the school, $320 - 225 = \mathbf{95}$ students are *not* involved in either band or sports.

Method II: Use Set Notation.

If B represents the set of students in the band, then $n(B)$ represents the number of students in set B, so $n(B) = 85$. Similarly, if S represents the set of all students on sports teams, $n(S) = 200$. Since 60 students participate in both activities, the number of students in the intersection of sets B and S is 60, so $n(B \cap S) = 60$. Hence:

$$n(B \cup S) = n(B) + n(S) - n(B \cap S)$$
$$= 85 \quad + 200 - 60$$
$$= 225$$

Since 225 students are involved in band or sports but not both, $320 - 225 =$ **95** students are *not* involved in either band or sports.

30. It is given that Ms. Brown plans to carpet part of her square living room floor, which is 20 feet by 20 feet. She wants to carpet a quarter-circle as shown in the accompanying diagram.

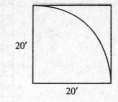

- Since the area A of a circle with radius r is given by the formula $A = \pi r^2$, the area of the quarter-circle to be carpeted is

$$A = \frac{1}{4} \times \pi \times (20)^2$$
$$= \frac{1}{4} \times \pi \times 400$$
$$= 100\pi \text{ square feet}$$

- The area of the square living room is 20×20 or 400 square feet.

- The area of the region that is uncarpeted is $400 - 100\pi$ square feet. Evaluate $400 - 100\pi$ using 3.14 as an approximation for π:

$$400 - 100\pi \approx 400 - 100(3.14)$$
$$\approx 400 - 314$$
$$\approx 86 \text{ square feet}$$

Thus, to the *nearest square foot*, **86** square feet of the living room will remain uncarpeted.

PART IV

31. It is given that video rental club *A* charges $12 for membership and $2 for each rented video, while video rental club *B* charges $4 for membership and $4 for each rented video. The accompanying graph represents the total cost of renting videos from Club *A*.

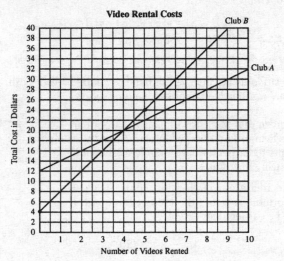

Notice that the line consists of the set of ordered pairs of the form
(*n* videos, total cost of renting *n* videos)

and that the *y*-intercept of the line at (0,12) is the cost of renting 0 videos at a cost of $12, which is the membership fee.

(a) To draw a line representing the total cost of renting videos from Club *B*, determine the *y*-intercept and the coordinates of another point on the line.

- Since the membership fee for Club *B* is $4, the *y*-intercept of line *B* is (0,4).

- Pick a convenient number of videos rented, say 5. The total cost of renting 5 videos from Club *B* is $4 + (5 × $4) = $4 + $20 = $24. Another point on the line is, therefore, (5,24).

- Plot (0,4) and (5,24), and connect these points with a straight line, as shown in the accompanying figure.

The line has a **slope of 4** and a ***y*-intercept of 4**.

(b) The line for Club *A* lies *below* the line for Club *B* when the number of videos rented is equal to or greater than 5. Hence, for **5 or more** video rentals, it is less expensive to belong to Club *A*.

32. It is given that Jed bought a generator that will run for 2 hours on a liter of gas. The gas tank on the generator is a rectangular prism with dimensions 20 cm by 15 cm by 10 cm as shown in the accompanying diagram.

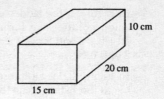

- The volume of the gas tank is the product of its length, width, and height: 20 cm × 15 cm × 10 cm or 3000 cm³.

- Since 1 liter = 1000 cm³, 3000 cm³ is equivalent to 3 liters.

- Because the gas tank runs 2 hours/liter, a full 3-liter gas tank will run 3 liters × 2 hours/liter or **6 hours**.

33. It is given that a clothing store offers a 50% discount at the end of each week that an item remains unsold. To show that after two weeks an unsold item is *not* 100% off, suppose the item originally costs $1.00.

- At the end of the first week, the dollar amount of the discount is 50% of $1.00 or $0.50. The sale price at the end of the first week is $1.00 − $0.50 (discount) = $0.50.

- At the end of the second week, the amount of the discount is 50% of the last sale price of $0.50 or $0.25. Therefore, the sale price at the end of the second week is $0.50 − $0.25 or $0.25.

- Since the saving at the end of the second week is $1.00 − $0.25 = $0.75, the correct percent of the discount is:

$$\frac{\text{total amount of saving}}{\text{original price}} \times 100 = \frac{0.75}{1.00} \times 100 = \mathbf{75\%}$$

34. It is given that, in Figure 1, a 10-foot ladder leans against a building so that the bottom of the ladder is 8 feet from the base of the building. When the ladder is moved 4 feet closer to the base of the building, as shown in Figure 2, the top of the ladder reaches further up the building.

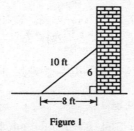

Figure 1

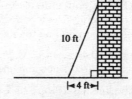

Figure 2

- In Figure 1, the ladder is the hypotenuse of a right triangle in which the lengths of the sides form a multiple of a 3–4–5 Pythagorean triple. Since $10 = 5 \times 2$ and $8 = 4 \times 2$, the vertical leg of the right triangle measures 3×2 or 6 feet.

- In Figure 2, let x represent the vertical leg of the right triangle. According to the Pythagorean theorem:

$$x^2 + 4^2 = 10^2$$
$$x^2 + 16 = 100$$
$$x^2 = 100 - 16$$
$$= 84$$
$$x = \sqrt{84}$$
$$\approx 9.17$$

- Since $9.17 - 6 = 3.17$, to the *nearest foot* the ladder in Figure 2 reaches **3 feet** further up the building than the ladder in Figure 1.

35. It is given that a corner is cut off a 5″ by 5″ square piece of paper. The cut is x inches from a corner as shown in the accompanying diagram.

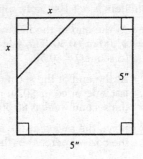

(a) The area of the original square is $5 \times 5 = 25$ square inches. Since the corner that is cut off is a right triangle, its area is equal to

$$\frac{1}{2} \times \text{base} \times \text{height} = \frac{1}{2} \cdot x \cdot x = \frac{x^2}{2}.$$

If A represents the area of the paper *after* the corner is removed, then A is the difference between the area of the original square and the area of the corner. Thus, $A = 25 - \dfrac{x^2}{2}$ is an equation that represents the area, A, of the paper after the corner is removed.

(b) If the area of the paper after the corner is removed is $\dfrac{7}{8}$ of the area of the original square, then:

$$\frac{7}{8}(25) = 25 - \frac{x^2}{2}$$

The smallest positive integer into which both denominators divide evenly is 8. Eliminate the fractions by multiplying each side of the equation by 8:

$$8 \times \frac{7}{8}(25) - 8\left(25 - \frac{x^2}{2}\right)$$

$$7 \times 25 = 8 \times 25 - \frac{8x^2}{2}$$

$$75 = 200 - 4x^2$$

$$175 - 200 = -4x^2$$

$$-25 = -4x^2$$

Divide each side by -4:

$$\frac{25}{4} = x^2$$

Or:

$$x^2 = \frac{25}{4}$$

$$x = \pm\sqrt{\frac{25}{4}} = \pm\frac{5}{2}$$

Since x represents a length, it cannot be negative. Hence, $x = \frac{5}{2}$ (or 2.5) inches.

The value of x is **2.5** inches.

Topic	Question Numbers	Number of Points	Your Points	Your Percentage
1. Numbers; Properties of real numbers; Percent	4	2		
2. Operations on Rat'l Numbers & Monomials	—	—		
3. Laws of Exponents for Integer Exponents; Scientific Notation	8	2		
4. Operations on Polynomials	5, 7	2 + 2 = 4		
5. Square Root; Operations with Radicals	19	2		
6. Evaluating Formulas & Alg. Expressions	—	—		
7. Solving Linear Eqs. & Inequalities	2, 13	2 + 2 = 4		
8. Solving Literal Eqs. & Formulas for a Particular Letter	—	—		
9. Alg. Operations (including factoring)	6	2		
10. Solving Quadratic Eqs.	—	—		
11. Coordinate Geometry (graphs of linear eqs.; slope; midpoint; distance)	11	2		
12. Systems of Linear Eqs. & Inequalities (alg. and graph. solutions)	31	4		
13. Mathematical Modeling Using: Eqs., Tables, Graphs of Linear Eqs.	27	3		
14. Linear-Quad. Systems (alg. and graph. solutions)	—	—		
15. Word Problems Requiring Arith. or Alg. Reasoning	16, 24, 28, 33, 35	2 + 2 + 3 + 4 + 4 = 15		
16. Areas, Perims., Circums., Vols. of Common Figures	30, 32	3 + 4 = 7		
17. Angle & Line Relationships (suppl., compl., vertical angles; parallel lines; congruence)	10, 15	2 + 2 = 4		
18. Ratio & Proportion (incl. similar polygons)	—	—		
19. Pythagorean Theorem	34	4		
20. Right Triangle Trig. & Indirect Measurement	25	2		
21. Logic (symbolic rep.; conditionals; logically equiv. statements; valid arguments)	12	2		
22. Probability (incl. tree diagrams & sample spaces)	20, 21	2 + 2 = 4		

Topic	Question Numbers	Number of Points	Your Points	Your Percentage
23. Counting Methods and Sets	14	$2 + 3 = 5$		
24. Permutations and Combinations	—	—		
25. Statistics (mean, percentiles, quartiles; freq. dist.; histograms; stem & leaf plots)	1, 18, 26	$2 + 2 + 3 = 7$		
26. Properties of Triangles & Parallelograms	9	2		
27. Transformations (reflections; translations; rotations; dilations)	3	2		
28. Symmetry	—	—		
29. Area & Transformations Using Coordinates	22, 23	$2 + 2 = 4$		
30. Locus & Constructions	17	2		
31. Dimensional Analysis	—	—		

Examination
Math A
June 1999

PART I

Answer all questions in this part. Each correct answer will receive two (2) credits. No partial credit will be allowed. Record your answers in the spaces provided. [40]

1 A fair coin is thrown in the air four times. If the coin lands with the head up on the first three tosses, what is the probability that the coin will land with the head up on the fourth toss?

  (1) 0   (3) $\frac{1}{8}$

  (2) $\frac{1}{16}$ (4) $\frac{1}{2}$ 1 _____

2 The statement "If x is divisible by 8, then it is divisible by 6" is false if x equals

  (1) 6   (3) 32

  (2) 14   (4) 48   2 _____

3 What is the image of point (2,5) under the translation that shifts (x,y) to $(x + 3, y - 2)$?

  (1) (0,3)   (3) (5,3)

  (2) (0,8)   (4) (5,8)   3 _____

4 The sum of $3x^2 + x + 8$ and $x^2 - 9$ can be expressed as

(1) $4x^2 + x - 1$ (3) $4x^4 + x - 1$

(2) $4x^2 + x - 17$ (4) $3x^4 + x - 1$ 4 _____

5 The direct distance between city A and city B is 200 miles. The direct distance between city B and city C is 300 miles. Which could be the direct distance between city C and city A?

(1) 50 miles   (3) 550 miles

(2) 350 miles   (4) 650 miles   5 _____

6 Expressed as a single fraction, what is $\dfrac{1}{x+1} + \dfrac{1}{x}$, $x \uparrow 0, -1$?

(1) $\dfrac{2x+3}{x^2+x}$ (3) $\dfrac{2}{2x+1}$

(2) $\dfrac{2x+1}{x^2+x}$ (4) $\dfrac{3}{x^2}$ 6 _____

7 How many different three-member teams can be formed from six students?

(1) 20   (3) 216

(2) 120   (4) 720   7 _____

8 If $x = -3$ and $y = 2$, which point on the accompanying graph represents $(-x, -y)$?

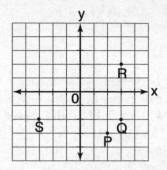

(1) P (3) R

(2) Q (4) S 8 _____

9 The larger root of the equation $(x + 4)(x - 3) = 0$ is

(1) -4 (3) 3

(2) -3 (4) 4 9 _____

10 Linda paid $48 for a jacket that was on sale for 25% of the original price. What was the original price of the jacket?

(1) $60   (3) $96

(2) $72   (4) $192   10 _____

11 The expression $2^3 \cdot 4^2$ is equivalent to

(1) 2^7 (3) 8^5

(2) 2^{12} (4) 8^6 11 _____

12 In the accompanying diagram of $\triangle ABC$, $\overline{AB}$ is extended to D, exterior angle CBD measure $145°$, and m$\angle C = 75$.

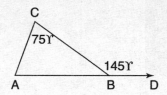

What is m$\angle CAB$?

(1) 35   (3) 110
(2) 70   (4) 220   12 _____

13 A total of $450 is divided into equal shares. If Kate receives four shares, Kevin receives three shares, and Anna receives the remaining two shares, how much money did Kevin receive?

(1) $100   (3) $200
(2) $150   (4) $250   13 _____

14 What is the diameter of a circle whose circumference is 5?

(1) $\dfrac{2.5}{\pi^2}$ (3) $\dfrac{5}{\pi^2}$

(2) $\dfrac{2.5}{\pi}$ (4) $\dfrac{5}{\pi}$ 14 _____

15 During a recent winter, the ratio of deer to foxes was 7 to 3 in one county of New York State. If there were 210 foxes in the county, what was the number of deer in the county?

(1) 90   (3) 280
(2) 147   (4) 490   15 _____

16 In the accompanying figure, *ACDH* and *BCEF* are rectangles, $AH = 2$, $GH = 3$, $GF = 4$, and $FE = 5$.

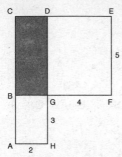

What is the area of *BCDG*?

(1) 6   (3) 10
(2) 8   (4) 20   16 _____

17 If $t^2 < t < \sqrt{t}$, then *t* could be

(1) $-\frac{1}{4}$ (3) $\frac{1}{4}$
(2) 0   (4) 4   17 _____

18 What is the slope of line ℓ shown in the accompanying diagram?

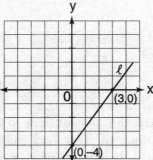

(1) $\frac{4}{3}$ (3) $-\frac{3}{4}$

(2) $\frac{3}{4}$ (4) $-\frac{4}{3}$ 18 _____

19 In a class of 50 students, 18 take music, 26 take art, and 2 take both art and music. How many students in the class are not enrolled in either music or art?

(1) 6   (3) 16
(2) 8   (4) 24   19 _____

20 The expression $\sqrt{27} + \sqrt{12}$ is equivalent to

(1) $5\sqrt{3}$ (3) $5\sqrt{6}$
(2) $13\sqrt{3}$ (4) $\sqrt{39}$ 20 _____

PART II

Answer all questions in this part. Each correct answer will receive two (2) credits. Clearly indicate the necessary steps, including appropriate formula substitutions, diagrams, graphs, charts, etc. For all questions in this part, a correct numerical answer with no work shown will receive only 1 credit. [10]

21 Draw all the symmetry lines on the accompanying figure.

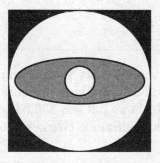

22 Shoe sizes and foot length are related by the formula $S = 3F - 24$, where S represents the shoe size and F represents the length of the foot, in inches.

  a Solve the formula for F

  b To the *nearest tenth of an inch,* how long is the foot of a person who wears a size $10\frac{1}{2}$ shoe?

23 Which number below is irrational?

$$\sqrt{\frac{4}{9}}, \quad \sqrt{20}, \quad \sqrt{121}$$

Why is the number you chose an irrational number?

_____

_____

_____

24 Simplify: $\dfrac{9x^2 - 15xy}{9x^2 - 25y^2}$

25 Sara's telephone service costs $21 per month plus $0.25 for each local call, and long-distance calls are extra. Last month, Sara's bill was $36.64, and it included $6.14 in long-distance charges. How many local calls did she make?

PART III

Answer all questions in this part. Each correct answer will receive three (3) credits. Clearly indicate the necessary steps, including appropriate formula substitutions, diagrams, graphs, charts, etc. For all questions in this part, a correct numerical answer with no work shown will receive only 1 credit. [15]

26 During a 45-minute lunch period, Albert (A) went running and Bill (B) walked for exercise. Their times and distances are shown in the accompanying graph. How much faster was Albert running than Bill was walking, in miles per hour?

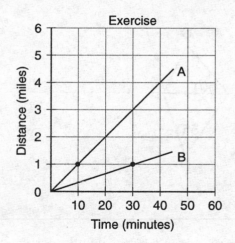

27 The dimensions of a brick, in inches, are 2 by 4 by 8. How many such bricks are needed to have a total volume of exactly 1 cubic foot?

28 A swimmer plans to swim at least 100 laps during a
6-day period. During this period, the swimmer will
increase the number of laps completed each day
by one lap. What is the *least* number of laps the
swimmer must complete on the first day?

29 The mean (average) weight of three dogs is 38
pounds. One of the dogs, Sparky, weighs 46
pounds. The other two dogs, Eddie and Sandy,
have the same weight. Find Eddie's weight.

30 In the accompanying diagram, $\triangle ABC$ and $\triangle ABD$
are isosceles triangles with m$\angle CAB = 50$ and
m$\angle BDA = 55$. If $AB = AC$ and $AB = BD$, what is
m$\angle CBD$?

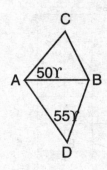

PART IV

Answer all questions in the part. Each correct answer will receive four (4) credits. Clearly indicate the necessary steps, including appropriate formula substitutions, diagrams, graphs, charts, etc. For all questions in this part, a correct numerical answer with no work shown will receive only 1 credit. [20]

31 A target shown in the accompanying diagram consists of three circles with the same center. The radii of the circles have lengths of 3 inches, 7 inches, and 9 inches.

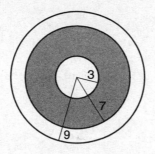

  a What is the area of the shaded region to the *nearest tenth of a square inch*?

  b To the *nearest percent*, what percent of the target is shaded?

32 A bookshelf contains six mysteries and three biographies. Two books are selected at random without replacement.

  a What is the probability that both books are mysteries?

  b What is the probability that one book is a mystery and the other is a biography?

33 The cross section of an attic is in the shape of an isosceles trapezoid, as shown in the accompanying figure. If the height of the attic is 9 feet, $BC = 12$ feet, and $AD = 28$ feet, find the length of $\overline{AB}$ to the *nearest foot*.

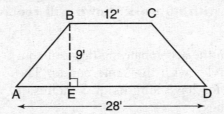

34 Joe is holding his kite string 3 feet above the ground, as shown in the accompanying diagram. The distance between his hand and a point directly under the kite is 95 feet. If the angle of elevation to the kite is 50°, find the height, h, of his kite, to the *nearest foot*.

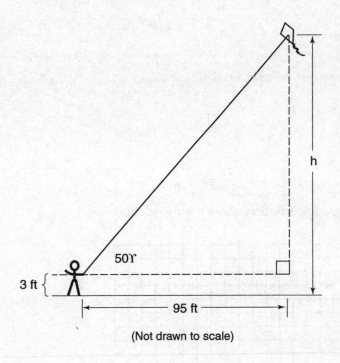

(Not drawn to scale)

35 Solve the following system of equations algebraically *or* graphically for *x* and *y*:

$$y = x^2 + 2x - 1$$
$$y = 3x + 5$$

For an algebraic solution, show your work here.

For a graphic solution, show your work here.

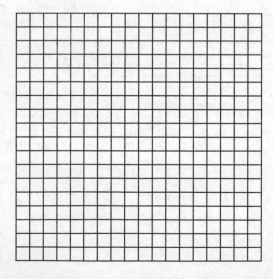

Answers
June 1999
Math A

Answer Key

PART I

1. (4)	**5.** (2)	**9.** (3)	**13.** (2)	**17.** (3)
2. (3)	**6.** (2)	**10.** (4)	**14.** (4)	**18.** (1)
3. (3)	**7.** (1)	**11.** (1)	**15.** (4)	**19.** (2)
4. (1)	**8.** (2)	**12.** (2)	**16.** (3)	**20.** (1)

PART II

21. See Answers Explained section.

22. a. $\dfrac{S+24}{3}$ or $\dfrac{S}{3}+8$

22. b. 11.5

23. $\sqrt{20}$ with explanation. (See Answers Explained section for explanation.)

24. $\dfrac{3x}{3x+5y}$

25. 38

PART III

26. 4   **27.** 27   **28.** 15   **29.** 34   **30.** 135

PART IV

31. a. 125.6 or 125.7
b. 49

32. a. $\dfrac{30}{72}$ **b.** $\dfrac{36}{72}$

33. 12

34. 116

35. (3,14) and (−2,−1)

Parts II–IV You are required to show how you arrived at your answers. For sample methods of solution, see Answers Explained section.

Answers Explained

PART I

1. The probability of obtaining a head when a fair coin is tossed is $\frac{1}{2}$. It is given that a fair coin lands with the head up on the first three tosses. Since the outcome of the fourth toss does not depend on the known outcomes of the preceding three tosses, the probability that the coin will land with the head up on the fourth toss is $\frac{1}{2}$.

The correct choice is **(4)**.

2. A conditional statement that has the form

"If <u>statement 1</u>, then <u>statement 2</u>"

is *false* only when statement 1 is true and statement 2 is false. Hence, the statement "If <u>x is divisible by 8</u>, then <u>it is divisible by 6</u>" is *false* for any value of x for which the first statement, "x is divisible by 8," is true and the second statement, "x is divisible by 6," is false. Thus, the correct answer choice will contain a number that is divisible by 8 but is *not* divisible by 6.

- Choice (1): Let $x = 6$. Since 6 is not divisible by 8, eliminate choice (1).

- Choice (2): Let $x = 14$. Since 14 is not divisible by 8, eliminate choice (2).

- Choice (3): Let $x = 32$. Because 32 is divisible by 8 but is not divisible by 6, choice (3) makes the given conditional a false statement.

- Choice (4): Let $x = 48$. Although 48 is divisible by 8, it is also divisible by 6, so eliminate choice (4).

The correct choice is **(3)**.

3. To find the image of point $(2,5)$ under the translation that shifts (x,y) to $(x + 3, y - 2)$, replace x with 2 and y with 5.

The given translation rule is: $\qquad (x,y) \rightarrow (x + 3, y - 2)$

Let $x = 2$ and $y = 5$: $\qquad (2,5) \rightarrow (2 + 3, 5 - 2) = (5,3)$

Thus, the image of $(2,5)$ under the given translation is $(5,3)$.

The correct choice is **(3)**.

4. To find the sum of $3x^2 + x + 8$ and $x^2 - 9$, write the second polynomial underneath the first polynomial, aligning like terms in the same vertical column. Then add like terms.

$$3x^2 + x + 8$$
$$\underline{\quad x^2 \quad - 9}$$
$$4x^2 + x - 1$$

The correct choice is **(1)**.

5. It is given that the direct distance between city A and city B is 200 miles, and the direct distance between city B and city C is 300 miles. The direct distance between city C and city A forms the third side of a triangle, as shown in the accompanying diagram.

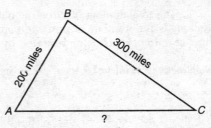

The length of each side of a triangle must be less than the sum of the lengths of the other two sides and greater than their difference. Since

$AB + BC = 200 + 300 = 500$ miles   and $BC - AB = 300 - 200 = 100$ miles,

the length of AC could be any number of miles that is between 100 and 500. Of the four answer choices, only 350 miles is between 100 miles and 500 miles, so this could be the direct distance between city C and city A.

The correct choice is **(2)**.

6. The given sum is: $\dfrac{1}{x+1} + \dfrac{1}{x} \quad (x \neq 0, -1)$

Find the least common denominator (LCD) of the two fractions. The LCD is the smallest expression into which both denominators will divide evenly. The LCD for $x + 1$ and x is $x(x + 1)$. Change the given fractions into equivalent fractions having the LCD as their denominators by multiplying the first fraction by 1 in the form of $\frac{x}{x}$ and multiplying the second fraction by 1

in the form of $\dfrac{x+1}{x+1}$:

$$\frac{x}{x}\left(\frac{1}{x+1}\right) + \left(\frac{x+1}{x+1}\right)\frac{1}{x}$$

$$\frac{x \cdot 1}{x(x+1)} + \frac{(x+1) \cdot 1}{x(x+1)}$$

$$\frac{x}{x(x+1)} + \frac{x+1}{x(x+1)}$$

$$\frac{x}{x^2+x} + \frac{x+1}{x^2+x}$$

Since the fractions now have the same
denominator, add them by writing the sum
of their numerators over their common denominator:

$$\frac{x+(x+1)}{x^2+x}$$

Combine like terms in the numerator:

$$\frac{2x+1}{x^2+x}$$

The correct choice is (2).

7. The total number of three-member teams that can be formed from six
students is the number of possible combinations of the six students taken
three at a time, which is represented by $_6C_3$. To find $_6C_3$, use a calculator or

evaluate the formula $_nC_r = \dfrac{n!}{r!(n-r)}$ using $n=6$ and $r=3$:

$$_6C_3 = \frac{6!}{3!(6-3)!}$$

$$= \frac{6 \times 5 \times 4 \times 3 \times 2 \times 1}{(3 \times 2 \times 1)(3 \times 2 \times 1)}$$

$$= \frac{\overset{1}{\cancel{6}} \times 5 \times 4 \times \cancel{3 \times 2 \times 1}}{(\cancel{3 \times 2 \times 1})(\cancel{3 \times 2 \times 1})}$$

$$= 5 \times 4$$

$$= 20$$

The correct choice is (1).

8. If $x = -3$ and $y = 2$, then $(-x,-y) =$
$(-(-3),-2) = (3,-2)$. To locate $(3,-2)$ on
the accompanying graph, place your pen
point at the origin. Since the x-coordinate
of $(3,-2)$ is *positive,* move the pen point
3 units to the *right,* along the positive
x-axis.

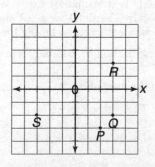

Since the y-coordinate of $(3,-2)$ is
negative, now move the pen point 2 units
vertically *down,* placing it on point Q,
which represents $(-x,-y)$.

The correct choice is (2).

9. If the product of two numbers is 0, then either number may be 0. Hence, if $(x + 4)(x - 3) = 0$, then:

$$x + 4 = 0 \quad \text{or} \quad x - 3 = 0$$
$$x = -4 \quad \text{or} \quad x = 3$$

Since 3 is greater than –4, the larger root of the equation is 3.

The correct choice is **(3)**.

10. It is given that Linda paid $48 for a jacket that was on sale for 25% of the original price. If x represents the original price of the jacket, then

$$0.25x = \$48$$
$$x = \$48 \div 0.25$$
$$= \$192$$

The correct choice is **(4)**.

11. The given expression is: $2^3 \cdot 4^2$
Substitute 2^2 for 4: $2^3 \cdot (2^2)^2$
Raise a power to a power by multiplying the
exponents: $2^3 \cdot 2^4$
Multiply powers of the same base by adding
their exponents: 2^{3+4} or 2^7

The correct choice is **(1)**.

12. In the accompanying diagram of $\triangle ABC$, $\overline{AB}$ is extended to D, exterior angle CBD measures 145°, and $m\angle C = 75$.

The measure of an exterior angle of a triangle is equal to the sum of the measures of the two nonadjacent interior angles:

$$m\angle CAB + m\angle C = m\angle CBD$$
$$m\angle CAB + 75 = 145$$
$$m\angle CAB = 145 - 75$$
$$= 70$$

The correct choice is **(2)**.

13. It is given that a total of $450 is divided into equal shares and that Kate receives four shares, Kevin receives three shares, and Anna receives the remaining two shares. The total number of shares is $4 + 3 + 2 = 9$, making the value of each share $\dfrac{\$450}{9} = \$50$.

Since Kevin has three shares, he receives $3 \times \$50 = \$150$.

The correct choice is **(2)**.

14. The circumference C of a circle is equal to π times the diameter D of the circle; that is, $C = \pi D$. If $C = 5$, then

$$5 = \pi D$$
$$D = \frac{5}{\pi}$$

The correct choice is **(4)**.

15. It is given that in a recent winter when the ratio of deer to foxes was 7 to 3, there were 210 foxes. If x represents the number of deer, then:

$$\frac{\text{deer}}{\text{foxes}} = \frac{7}{3} = \frac{x}{210}$$

In a proportion, the product of the means is equal to the product of the extremes (cross-multiply):

$$3x = 7(210)$$
$$= 1470$$
$$x = \frac{1470}{3} = 490$$

The correct choice is **(4)**.

16. It is given that, in the accompanying diagram, $ACDH$ and $BCEF$ are rectangles, $AH = 2$, $GH = 3$, $GF = 4$, and $FE = 5$.

Opposite sides of a rectangle are equal in length, so

$$CD = AH = 2$$

and

$$BC = FE = 5$$

Since $BCDG$ is also a rectangle,

area of $BCDG = CD \times BC$

$$= 2 \times 5$$
$$= 10$$

The correct choice is **(3)**.

17. To find the answer choice that contains a value of t that makes $t^2 < t < \sqrt{t}$ a true statement, consider each choice in turn.

- Choice (1): Let $t = -\dfrac{1}{4}$. Since $\sqrt{-\dfrac{1}{4}}$ is not a real number, eliminate choice (1).

- Choice (2): Let $t = 0$. Since it is not true that $0^2 < 0 < \sqrt{0}$, eliminate choice (2).

- Choice (3): Let $t = \dfrac{1}{4}$. Determine whether $\left(\dfrac{1}{4}\right)^2 < \dfrac{1}{4} < \sqrt{\dfrac{1}{4}}$ is true.

  Since $\left(\dfrac{1}{4}\right)^2 = \dfrac{1}{16}$ and $\sqrt{\dfrac{1}{4}} = \dfrac{1}{2}$, the inequality $\dfrac{1}{16} < \dfrac{1}{4} < \dfrac{1}{2}$

  is true, so t could be equal to $\dfrac{1}{4}$.

- Choice (4): Let $t = 4$. Since $4^2 \,(=16) < 4 < \sqrt{4}\,(= 2)$ is false, eliminate choice (4).

The correct choice is **(3)**.

18. The slope of a line is a measure of its steepness and can be determined by writing the difference of the y-coordinates of any two points on the line over the difference of the x-coordinates of the same two points.

Using points $(0,-4)$ and $(3,0)$, calculate the slope of line ℓ in the accompanying diagram by subtracting the y- and x-coordinates of the second point from the corresponding coordinates of the first point:

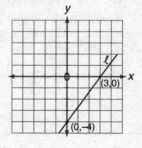

$$\text{slope} = \frac{\text{difference in } y\text{-coordinates}}{\text{difference in } x\text{-coordinates}}$$

$$= \frac{-4 - 0}{0 - 3} = \frac{-4}{-3} = \frac{4}{3}$$

The correct choice is **(1)**.

19. It is given that, in a class of 50 students, 18 take music, 26 take art, and 2 take both art and music, as represented in the accompanying Venn diagram. The 2 students taking both art and music are counted in the 18 taking music and are counted again in the 26 taking art. Hence, of the 50 students in the class, the number of students taking art or music or both art and music is $18 + 26 - 2$ or 42.

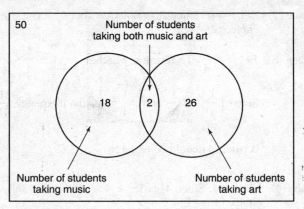

Thus, $50 - 42$ or 8 students are not enrolled in either music or art.

The correct choice is **(2)**.

20. The given expression is:

$$\sqrt{27} + \sqrt{12}$$

Rewrite each number underneath the radical sign as the product of two integers one of which is the greatest perfect square factor of the number:

$$\sqrt{9 \cdot 3} + \sqrt{4 \cdot 3}$$

Write the radical over each factor:

$$\sqrt{9} \cdot \sqrt{3} + \sqrt{4} \cdot \sqrt{3}$$

Evaluate the square roots of the perfect squares:

$$3\sqrt{3} + 2\sqrt{3}$$

Combine:

$$5\sqrt{3}$$

The correct choice is **(1)**.

PART II

21. As shown in the accompanying figure, there are **two lines of symmetry: one horizontal and one vertical**.

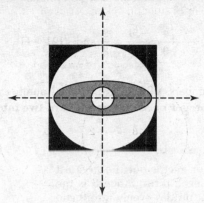

22. It is given that shoe sizes and foot length are related by the formula $S = 3F - 24$, where S represents the shoe size and F represents the length of the foot, in inches.

a. To solve the formula for F, isolate F.

Add 24 to each side of the equation:
$$S + 24 = 3F - 24 + 24$$
$$= 3F$$

Divide each side of the equation by 3:
$$\frac{S + 24}{3} = F$$

or, equivalently,
$$F = \frac{S + 24}{3} \quad \text{or} \quad \frac{S}{3} + 8$$

b. To find the length of the foot of a person who wears a size $10\frac{1}{2}$ shoe, substitute $10\frac{1}{2}$ for S in the formula $F = \frac{S + 24}{3}$:

$$F = \frac{10\frac{1}{2} + 24}{3}$$
$$= \frac{10.5 + 24}{3}$$
$$= \frac{34.5}{3}$$
$$= 11.5$$

The length of the foot is, to the *nearest tenth of an inch*, **11.5**.

23. Of the three given numbers, $\sqrt{\dfrac{4}{9}}, \sqrt{20}$, and $\sqrt{121}$, the radicals in the first and the last of these numbers can be eliminated, so they are rational numbers:

$$\sqrt{\frac{4}{9}} = \frac{\sqrt{4}}{\sqrt{9}} = \frac{2}{3} \text{ and } \sqrt{121} = 11$$

Of the three numbers, $\sqrt{20}$ is an irrational number since **it cannot be expressed as an integer or as the quotient of two integers**.

24. The given fraction is:

$$\frac{9x^2 - 15xy}{9x^2 - 25y^2}$$

In the numerator, factor out the greatest common factor of each term. Since 3 is the greatest integer that divides evenly into 9 and 15 and x is the greatest literal factor common to both x^2 and xy, the greatest common factor is $3x$:

$$\frac{3x(3x - 5y)}{9x^2 - 25y^2}$$

The denominator is the difference between two squares since it has the form $A^2 - B^2$, where $A = 3x$ and $B = 5y$. The difference between two squares can be factored as the product of the sum and difference of the two terms that are being squared; that is, $A^2 - B^2 = (A + B)(A - B)$, so

$$9x^2 - 25y^2 = (3x)^2 - (5y)^2$$

$$= (3x + 5y)(3x - 5y): \quad \frac{3x\cancel{(3x-5y)}^{\,1}}{(3x + 5y)\cancel{(3x-5y)}}$$

Divide out any factor that is common to both the numerator and the denominator since their quotient is 1:

$$\frac{3x}{3x + 5y}$$

The fraction in simplest form is $\dfrac{3x}{3x + 5y}$.

25. It is given that Sara's telephone service costs $21 per month plus $0.25 for each local call, and long-distance calls are extra. Last month, Sara's bill of $36.64 included $6.14 in long-distance charges.

Sara's bill without the long-distance charges was $36.64 − $6.14 = $30.50.

If x represents the number of local calls Sara made, then

$$\$21 + \$0.25x = \$30.50$$

$$\$0.25x = \$30.50 - \$21$$

$$= \$9.50$$

$$x = \frac{\$9.50}{\$0.25} = 38$$

Sara made **38** local calls.

PART III

26. It is given that, in the accompanying graph, line A shows the time and distance for Albert when he went running during a 45-minute lunch period. Also, line B shows the time and distance for Bill when he walked for exercise during the same 45-minute lunch period.

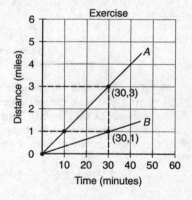

- The slope of line A represents the rate at which Albert was running. Pick any convenient point on line A. For example, when time = 30 minutes, distance = 3 miles. The problem asks for the rate in miles

per *hour*, so use $\frac{1}{2}$ hour instead of 30 minutes. Also, since the origin is on line A, use (0,0) as the other point in the slope calculation:

$$\text{slope of line } A = \frac{\text{change in distance}}{\text{change in time}} = \frac{3 - 0 \text{ miles}}{\frac{1}{2} - 0 \text{ hours}}$$

$$= 3 \div \tfrac{1}{2} = \ 6 \text{ miles per hour}$$

• The slope of line B represents the rate at which Bill was walking. Pick any convenient point on line B. For example, when time = 30 minutes = $\frac{1}{2}$ hour, distance = 1 mile. Since the origin is also on line B, use (0,0) as the other point in the slope calculation:

$$\text{slope of line B} = \frac{\text{change in distance}}{\text{change in time}} = \frac{1 - 0 \text{ miles}}{\frac{1}{2} - 0 \text{ hours}}$$
$$= 1 \div \frac{1}{2} = 2 \text{ miles per hour}$$

Since $6 - 2 = 4$, Albert was running **4** miles per hour faster than Bill was walking.

27. If the dimensions of a brick, in inches, are 2 by 4 by 8, then the volume of one brick, in cubic inches, is $2 \times 4 \times 8$ or 64. A volume of 1 cubic foot is equivalent, in cubic inches, to $12 \times 12 \times 12 = 1728$.

Hence, the number of bricks needed to have a volume of 1728 cubic inches (1 cubic foot) is

$$\frac{1728 \text{ cubic inches}}{64 \text{ cubic inches}} = \mathbf{27}.$$

28. It is given that a swimmer plans to swim at least 100 laps during a 6-day period and that the swimmer will increase the number of laps completed each day by one lap. If x represents the least number of laps the swimmer must complete on the first day, then $x + 1, x + 2, x + 3, x + 4$, and $x + 5$ represent the number of laps completed on the next five consecutive days. Hence:

$$x + (x + 1) + (x + 2) + (x + 3) + (x + 4) + (x + 5) \geq 100$$
$$6x + 15 \geq 100$$
$$6x \geq 85$$
$$x \geq \frac{85}{6}$$
$$x \geq 14.16$$

Since x must be an integer, the least value of x is 15. Hence, the *least* number of laps the swimmer must complete on the first day is **15**.

29. It is given that the mean (average) weight of three dogs is 38 pounds. One of the dogs, Sparky, weighs 46 pounds, and the other two dogs, Eddie and Sandy, have the same weight.

Let x = Eddie's weight. Then:

$$\frac{\text{sum of three weights}}{3} = \text{mean weight}$$

$$\frac{x + x + 46}{3} = 38$$

$$\frac{2x + 46}{3} = 38$$

Multiply each side of the equation by 3:

$$3\left(\frac{2x + 46}{3}\right) = 3(38)$$

$$2x + 46 = 114$$

$$2x = 114 - 46$$

$$x = \frac{68}{2} = 34$$

Eddie's weight is **34** pounds.

30. It is given that, in the accompanying diagram, $\triangle ABC$ and $\triangle ABD$ are isosceles triangles with m$\angle CAB$ = 50, m$\angle BDA$ = 55, $AB = AC$, and $AB = AD$. To find the measure of m$\angle CBD$, find the measures of angles CBA and DBA and then add these measures together.

- Since the sum of the measures of the three angles of a triangle is 180, in $\triangle ABC$

$$\text{m}\angle CBA + \text{m}\angle C + 50 = 180$$

$$\text{m}\angle CBA + \text{m}\angle C = 130$$

In isosceles triangle ABC, the angles opposite legs AB and AC have equal measures, so

$$\text{m}\angle CBA = \text{m}\angle C = \frac{1}{2}(130) = 65$$

- In isosceles triangle ABD, the angles opposite legs AB and BD have equal measures, so

$$\text{m}\angle BAD = \text{m}\angle BDA = 55$$

Since the sum of the measures of the three angles of a triangle is 180°:

$$\text{m}\angle DBA + \text{m}\angle BAD + \text{m}\angle BDA = 180$$

$$\text{m}\angle DBA + 55 + 55 = 180$$

$$\text{m}\angle DBA + 110 = 180$$

$$\text{m}\angle DBA = 180 - 110 = 70$$

Hence:

$$m\angle CBD = m\angle CBA + m\angle DBA$$
$$= 65 + 70$$
$$= 135$$

The degree measure of $\angle CBD$ is **135**.

PART IV

31. It is given that the target shown in the accompanying diagram consists of three circles with the same center and radii with lengths of 3 inches, 7 inches, and 9 inches. The formula for the area of a circle is $\pi \times r^2$, where r is the length of the radius of the circle.

a. The area of the shaded region is the difference between the area of the circle with the 7-inch radius and the area of the circle with the 3-inch radius.

The area of the circle with a 7-inch radius is $\pi \times 7^2 = 49\pi$, and the area of the circle with the 3-inch radius is $\pi \times 3^2 = 9\pi$.

Hence, the area of the shaded region is $49\pi - 9\pi = 40\pi$.

The question asks for the area to the nearest tenth of a square inch, so evaluate 40π, using 3.14 as an approximation for π. Since $40\pi = 40 \times 3.14 = 125.6$, the area to the *nearest tenth of a square inch* is **125.6**. [Note: If you use a more precise approximation for π, the correct answer is **125.7**].

b. The area of the target is the area of the circle with a 9-inch radius or $\pi \times 9^2 = 81\pi$. Hence:

$$\text{percent of target shaded } = \frac{\text{area of shaded region}}{\text{area of target}} \times 100\%$$

$$= \frac{40\pi}{81\pi} \times 100\%$$

$$\approx 0.4938 \times 100\%$$

$$\approx 49.38\%$$

The percent of the target that is shaded, correct to the *nearest percent*, is **49**.

32. It is given that two books are selected at random without replacement from a bookshelf that contains six mysteries and three biographies.

a. To find the probability that both books are mysteries, determine the probabilities that the first book selected will be a mystery and the second book selected will be a mystery.

- Of the nine (= 6 + 3) books on the bookshelf, six are mysteries. Hence, the probability that the first book selected will be a mystery is $\frac{6}{9}$.

- Since there is no replacement, eight books are left on the bookshelf. Assume that the first book selected is a mystery; then five of the eight remaining books are mysteries. Hence, the probability that the second book selected will be a mystery is $\frac{5}{8}$.

- The probability that the first book selected will be a mystery *and* the second book selected will be a mystery is equal to the *product* of the individual probabilities of selecting books of these types. Hence:

$$P(\text{mystery, mystery}) = P(\text{mystery}) \times P(\text{mystery})$$

$$= \frac{6}{9} \times \frac{5}{8}$$

$$= \frac{30}{72}$$

The probability that both books are mysteries is $\dfrac{30}{72}$.

b. The order in which the two books are selected matters. The probability that one book will be a mystery and the other book will be a biography is the sum of the probabilities of selecting the two types of books in either order.

- The probability that the first book selected will be a mystery is $\frac{6}{9}$. Of the eight books left on the bookshelf, three are biographies, so the probability that the second book selected will be a biography is $\frac{3}{8}$. Thus, the probability of selecting a mystery first and then a biography, without replacement, is $\frac{6}{9} \times \frac{3}{8} = \frac{18}{72}$.

- The probability that the first book selected will be a biography is $\frac{3}{9}$. Of the eight books left on the bookshelf, six are mysteries, so the probability that the second book selected will be a mystery is $\frac{6}{8}$. Thus, the probability of selecting a biography first and then a mystery, without replacement, is $\frac{3}{9} \times \frac{6}{8} = \frac{18}{72}$.

Thus:

$$P(\text{mystery and biography}) = P(\text{mystery, biography}) + P(\text{biography, mystery})$$

$$= \frac{18}{72} + \frac{18}{72}$$

$$= \frac{36}{72}$$

The probability that one book is a mystery and the other is a biography is $\frac{36}{72}$.

33. It is given that, in the accompanying diagram of isosceles trapezoid $ABCD$, height $BE = 9$ feet, $BC = 12$ feet, and $AD = 28$ feet.

To find the length of $\overline{AB}$, first drop an altitude from C, intersecting AD at F, thus forming rectangle $EBCF$.

- Since opposite sides of a rectangle have the same length, $EF = BC = 12$ and $CF = BE = 9$.

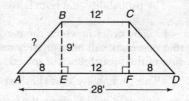

- Right triangle $AEB \cong$ right triangle DFC since the hypotenuse $(\overline{AB} \cong \overline{CD})$ and a leg $(\overline{BE} \cong \overline{CF})$ of one right triangle are congruent to the corresponding parts of the other right triangle. Thus, $AE = DF$.

- Since $AE + 12 + DF = 28$, then $AE + DF = 28 - 12 = 16$. Also, because $AE = DF$:

$$AE + AE = 16$$

$$AE = \tfrac{1}{2}(16) = 8$$

In right triangle $\triangle AEB$, find hypotenuse AB using the Pythagorean theorem:

$$(AB)^2 = (AE)^2 + (BE)^2$$

$$= 8^2 + 9^2$$

$$= 64 + 81$$

$$= 145$$

$$AB = \sqrt{145} \approx 12.02$$

The length of $\overline{AB}$ to the *nearest foot* is **12.**

34.

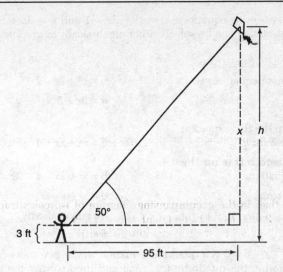

It is given that, in the accompanying diagram, a kite string is being held 3 feet above the ground. The distance between the holder's hand and a point directly under the kite is 95 feet, the angle of elevation to the kite is 50°, and h represents the height, in feet, of the kite.

Let x represent the length of the side of the right triangle that is opposite the 50° angle. Then:

$$\tan 50° = \frac{\text{side opposite 50° angle}}{\text{side adjacent to 50° angle}}$$

$$= \frac{x}{95}$$

Use your calculator to obtain $\tan 50° \approx 1.1918$:

$$1.1918 = \frac{x}{95}$$

Multiply each side of the equation by 95:

$$95 \times 1.1918 = 95 \times \frac{x}{95}$$

$$113.221 = x$$

Since the kite is being held 3 feet above the ground:

$$h = 3 + 113.221 = 116.221$$

Hence, the height of the kite, to the *nearest foot,* is **116** feet.

35. The given system of equations, $y = x^2 + 2x - 1$ and $y = 3x + 5$, is a linear-quadratic system that can be solved either algebraically or graphically.

<u>METHOD 1: ALGEBRAIC SOLUTION</u>

The given system of equations is:
$$y = x^2 + 2x - 1$$
$$y = 3x + 5$$

Eliminate y from the first equation by substituting $3x + 5$ for y:
$$3x + 5 = x^2 + 2x - 1$$

Collect all nonzero terms on the same side of the equation:
$$0 = x^2 + 2x - 1 - 3x - 5$$

Combine like terms:
$$0 = x^2 - x - 6$$

or equivalently,
$$x^2 - x - 6 = 0$$

The equation $x^2 - x - 6 = 0$ is a quadratic equation whose two roots can be determined by factoring the quadratic trinomial and then solving the equations that result from setting each factor equal to 0.

The binomials factors of $x^2 - x - 6$ have the form $(x + ?)(x + ?)$. The missing numbers are the two integers whose product is -6, the last term of $x^2 - x - 6$, and whose sum is -1, the coefficient of the x-term in $x^2 - x - 6$.

Since $(-3)(+2) = -6$ and $(-3) + (+2) = -1$, the two integers that satisfy these conditions are -3 and $+2$:
$$(x - 3)(x + 2) = 0$$

If the product of two numbers is 0, then at least one of these numbers is 0:
$$x - 3 = 0 \quad \text{or} \quad x + 2 = 0$$
$$x = 3 \quad \text{or} \quad x = -2$$

Find the corresponding values for y by substituting 3 and -2 for x in either of the two original equations.

- If $x = 3$, then $y = 3x + 5 = 3(3) + 5 = 9 + 5 = 14$. Hence, one solution is **(3,14)**.

- If $x = -2$, then $y = 3x + 5 = 3(-2) + 5 = -6 + 5 = -1$. Hence, the other solution is **(−2,−1)**.

<u>METHOD 2: GRAPHICAL SOLUTION</u>

To solve a linear-quadratic system of equations graphically, graph each equation on the same set of axes. The points of intersection of the two graphs, if any, represent the solutions to the system of equations.

• To graph $y = 3x + 5$, plot any two points that satisfy the equation and then use a straightedge to draw the line that connects these points. If $x = 0$, then $y = 5$, so $(0,5)$ satisfies the equation. Also, if $x = 2$, then $y = 3(2) + 5 = 6 + 5 = 11$, so $(2,11)$ satisfies the equation. Plot $(0,5)$ and $(2,11)$. Then draw a straight line through these points, as shown in the accompanying diagram.

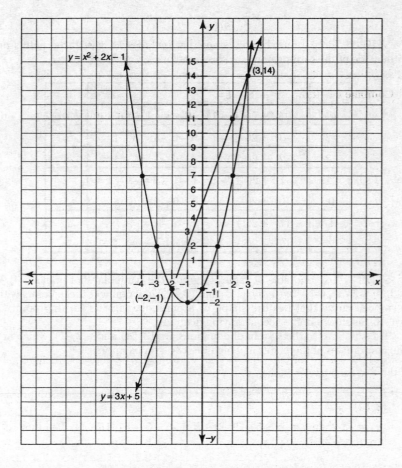

• Since $y = x^2 + 2x - 1$ is an equation in which only variable x is squared, its graph is a *parabola* with a vertical axis of symmetry. To draw the graph of a parabola, follow these steps:

STEP 1. Find the x-coordinate of the turning point of the parabola. In general, the x-coordinate of the turning point of a parabola whose equation has the form $y = ax^2 + bx + c$ is $x = -\dfrac{b}{2a}$. For the equation $y = x^2 + 2x -1$, $a = 1$ and $b = 2$, so the x-coordinate of the turning point is

$$x = -\frac{2}{2(1)} = -\frac{2}{2} = -1.$$

STEP 2. Prepare a table of values that includes at least three points on either side of the turning point, as shown below.

x	$x^2 + 2x - 1 =$	y
-4	$(-4)^2 + 2(-4) - 1 = 16 - 8 - 1 = 8 - 1 =$	7
-3	$(-3)^2 + 2(-3) - 1 = 9 - 6 - 1 = 3 - 1 =$	2
-2	$(-2)^2 + 2(-2) - 1 = 4 - 4 - 1 = 0 - 1 =$	-1
-1	$(-1)^2 + 2(-1) - 1 = 1 - 1 - 2 = -1 - 1 =$	-2
0	$(0)^2 + 2(0) - 1 = 0 + 0 - 1 =$	-1
1	$(1)^2 + 2(1) - 1 = 1 + 2 - 1 =$	2
2	$(2)^2 + 2(2) - 1 = 4 + 4 - 1 = 8 - 1 =$	7

STEP 3. Use the table of values to graph the parabola. Plot points $(-4,7)$, $(-3,2)$, $(-2,-1)$, $(-1,-2)$, $(0,-1)$, $(1,2)$ and $(2,7)$. Then connect these points with a smooth curve that has the shape of a parabola with its turning point at $(-1,-2)$, as shown in the accompanying diagram.

The solutions to the system of equations are **(3,14)** and **(-2,-1)**, which are the coordinates of the points at which the two graphs intersect.

Topic	Question Numbers	Number of Points	Your Points	Your Percentage
1. Numbers; Properties of real numbers; Percent	17, 23	2 + 2 = 4		
2. Operations on Rat'l Numbers & Monomials	—	—		
3. Laws of Exponents for Integer Exponents; Scientific Notation	11	2		
4. Operations on Polynomials	4	2		
5. Square Root; Operations with Radicals	20	2		
6. Evaluating Formulas & Alg. Expressions	22b	1		
7. Solving Linear Eqs. & Inequalities	—	—		
8. Solving Literal Eqs. & Formulas for a Particular Letter	22a	1		
9. Alg. Operations (including factoring)	6, 24	2 + 2 = 4		
10. Solving Quadratic Eqs.	9	2		
11. Coordinate Geometry (graphs of linear eqs.; slope; midpoint; distance)	8, 18	2 + 2 = 4		
12. Systems of Linear Eqs. & Inequalities (alg. and graph. solutions)	—	—		
13. Mathematical Modeling Using: Eqs., Tables, Graphs of Linear Eqs.	26	3		
14. Linear-Quad. Systems (alg. and graph. solutions)	35	4		
15. Word Problems Requiring Arith. or or Alg. Reasoning	10, 13, 25, 28	2 + 2 + 2 + 3 = 9		
16. Areas, Perims., Circums., Vols. of Common Figures	14, 16, 27, 31	2 + 2 + 3 + 4 = 11		
17. Angle & Line Relationships (suppl., compl., vertical angles; parallel lines; congruence)	—	—		
18. Ratio & Proportion (incl. similar polygons)	15	2		
19. Pythagorean Theorem	—	—		
20. Right Triangle Trig. & Indirect Measurement	34	4		
21. Logic (symbolic rep.; conditionals; logically equiv. statements; valid arguments)	2	2		
22. Probability (incl. tree diagrams & sample spaces)	1, 32	2 + 4 = 6		

Topic	Question Numbers	Number of Points	Your Points	Your Percentage
23. Counting Methods and Sets	19	2		
24. Permutations and Combinations	7	2		
25. Statistics (mean, percentiles, quartiles; freq. dist.; histograms; stem & leaf plots)	5, 29	2 + 3 = 5		
26. Properties of Triangles & Parallelograms	12, 30, 33	2 + 3 + 4 = 9		
27. Transformations (reflections; translations; rotations; dilations)	3	2		
28. Symmetry	21	2		
29. Area & Transformations Using Coordinates	—	—		
30. Locus & Constructions	—	—		
31. Dimensional Analysis	—	—		

HOW TO CONVERT YOUR RAW SCORE TO YOUR MATH A REGENTS EXAMINATION SCORE

Below is the conversion chart that must be used to determine your final score on the June 1999 Regents Examination in Mathematics A. To find your final exam score, locate in the column labeled "Raw Score" the total number of points you scored out of a possible 85 points. Then locate in the adjacent column to the right the scaled score that corresponds to your raw score. The scaled score is your final Mathematics A Regents Examination score.

Regents Examination in Mathematics A—June 1999
Chart for Converting Total Test Raw Scores to
Final Examination Scores (Scaled Scores)

Raw Score	Scaled Score	Raw Score	Scaled Score	Raw Score	Scaled Score
85	100	56	78	27	47
84	99	55	77	26	46
83	99	54	76	25	45
82	99	53	75	24	44
81	99	52	74	23	43
80	99	51	73	22	42
79	98	50	72	21	40
78	97	49	71	20	39
77	96	48	70	19	38
76	95	47	69	18	37
75	94	46	68	17	36
74	94	45	67	16	35
73	93	44	66	15	34
72	92	43	65	14	33
71	91	42	64	13	32
70	90	41	63	12	31
69	89	40	62	11	30
68	88	39	61	10	29
67	87	38	60	9	28
66	87	37	59	8	27
65	86	36	58	7	26
64	85	35	56	6	25
63	84	34	55	5	24
62	83	33	54	4	23
61	82	32	53	3	22
60	81	31	52	2	21
59	80	30	51	1	10
58	79	29	49	0	0
57	78	28	48		

Examination
August 1999
Math A

PART I

Answer all questions in this part. Each correct answer will receive two (2) credits. No partial credit will be allowed. Record your answers in the spaces provided. [40]

1 A roll of candy is shown in the accompanying diagram.

The shape of the candy is best described as a

(1) rectangular solid   (3) cone
(2) pyramid   (4) cylinder   1 _____

2 The expression $\sqrt{50}$ can be simplified to

(1) $5\sqrt{2}$ (3) $2\sqrt{25}$
(2) $5\sqrt{10}$ (4) $25\sqrt{2}$ 2 _____

3 The transformation of $\triangle ABC$ to $\triangle AB'C'$ is shown in the accompanying diagram.

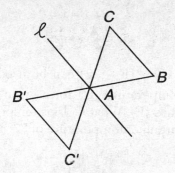

This transformation is an example of a

(1) line reflection in line ℓ
(2) rotation about point A
(3) dilation
(4) translation

3 ____

4 Which expression is equivalent to 6.02×10^{23}?

(1) 0.602×10^{21} (3) 602×10^{21}
(2) 60.2×10^{21} (4) 6020×10^{21}

4 ____

5 The Pentagon building in Washington, D.C., is shaped like a regular pentagon. If the length of one side of the Pentagon is represented by $n + 2$, its perimeter would be represented by

(1) $5n + 10$ (3) $n + 10$
(2) $5n + 2$ (4) $10n$

5 ____

6 The product of $4x^2y$ and $2xy^3$ is

(1) $8x^2y^3$ (3) $8x^3y^4$
(2) $8x^3y^3$ (4) $8x^2y^4$

6 ____

7 Which equation is an illustration of the additive identity property?

(1) $x \cdot 1 = x$ (3) $x - x = 0$

(2) $x + 0 = x$ (4) $x \cdot \frac{1}{x} = 1$ 7 ___

8 The formula $C = \frac{5}{9}(F - 32)$ can be used to find the Celsius temperature (C) for a given Fahrenheit temperature (F). What Celsius temperature is equal to a Fahrenheit temperature of 77°?

(1) 8°   (3) 45°

(2) 25°   (4) 171°   8 ___

9 In the accompanying diagram of rectangle $ABCD$, m$\angle BAC = 3x + 4$ and m$\angle ACD = x + 28$.

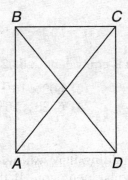

What is m$\angle CAD$?

(1) 12   (3) 40

(2) 37   (4) 50   9 ___

10 On June 17, the temperature in New York City ranged from 90° to 99°, while the temperature in Niagara Falls ranged from 60° to 69°. The difference in the temperatures in these two cities must be between

(1) 20° and 30°   (3) 25° and 35°
(2) 20° and 40°   (4) 30° and 40°   10 _____

11 Which expression is equivalent to $\dfrac{a}{x} + \dfrac{b}{2x}$?

(1) $\dfrac{2a+b}{2x}$ (3) $\dfrac{a+b}{3x}$

(2) $\dfrac{2a+b}{x}$ (4) $\dfrac{a+b}{2x}$ 11 _____

12 What is true about the statement "If two angles are right angles, the angles have equal measure" and its converse "If two angles have equal measure then the two angles are right angles"?

(1) The statement is true but its converse is false.
(2) The statement is false but its converse is true.
(3) Both the statement and its converse are false.
(4) Both the statement and its converse are true.   12 _____

13 If 6 and x have the same mean (average) as 2, 4, and 24, what is the value of x?

(1) 5   (3) 14
(2) 10   (4) 36   13 _____

14 In a hockey league, 87 players play on seven different teams. Each team has at least 12 players. What
is the largest possible number of players on any
one team?

(1) 13   (3) 15
(2) 14   (4) 21   14 _____

15 In the accompanying diagram of equilateral triangle *ABC*, $DE = 5$ and $\overline{DE} \parallel \overline{AB}$.

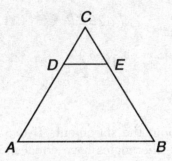

If *AB* is three times as long as *DE*, what is the
perimeter of quadrilateral *ABED*?

(1) 20   (3) 35
(2) 30   (4) 40   15 _____

16 At a concert, $720 was collected for hot dogs, ham-
burgers, and soft drinks. All three items sold for $1.00
each. Three times as many soft drinks were sold as
hamburgers. Twice as many hot dogs were sold as
hamburgers. The number of soft drinks sold was

(1) 120   (3) 360
(2) 240   (4) 480   16 _____

17 How many different 6-letter arrangements can be
formed using the letters in the word "ABSENT," if
each letter is used only once?

(1) 6   (3) 720
(2) 36   (4) 46,656   17 _____

18 The ratio of the corresponding sides of two similar
squares is 1 to 3. What is the ratio of the area of
the smaller square to the area of the larger square?

(1) $1:\sqrt{3}$ (3) 1:6
(2) 1:3   (4) 1:9   18 _____

19 What is the slope of the line whose equation is
$3x - 4y - 16 = 0$?

(1) $\frac{3}{4}$ (3) 3

(2) $\frac{4}{3}$ (4) −4   19 _____

20 What is the perimeter of an equilateral triangle
whose height is $2\sqrt{3}$?

(1) 6   (3) $6\sqrt{3}$

(2) 12   (4) $12\sqrt{3}$ 20 _____

PART II

Answer all questions in this part. Each correct answer will receive two (2) credits. Clearly indicate the necessary steps, including appropriate formula substitutions, diagrams, graphs, charts, etc. For all questions in this part, a correct numerical answer with no work shown will receive only 1 credit. [10]

21 Solve for x: $2(x - 3) = 1.2 - x$

22 The Grimaldis have three children born in different years.

a Draw a tree diagram or list a sample space to show all the possible arrangements of boy and girl children in the Grimaldi family.

b Using your information from part *a*, what is the probability that the Grimaldis have three boys?

23 Paloma has 3 jackets, 6 scarves, and 4 hats. Determine the number of different outfits consisting of a jacket, a scarf, and a hat that Paloma can wear.

24 In a recent poll, 600 people were asked whether they liked Chinese food. A circle graph was constructed to show the results. The central angles for two of the three sectors are shown in the accompanying diagram. How many people had no opinion?

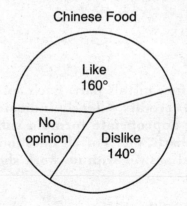

Chinese Food

25 Maria's backyard has two trees that are 40 feet apart, as shown in the accompanying diagram. She wants to place lampposts so that the posts are 30 feet from both of the trees. Draw a sketch to show where the lampposts could be placed in relation to the trees. How many locations for the lampposts are possible?

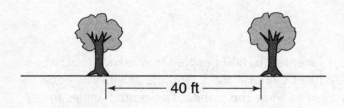

PART III

Answer all questions in this part. Each correct answer will receive three (3) credits. Clearly indicate the necessary steps, including appropriate formula substitutions, diagrams, graphs, charts, etc. For all questions in this part, a correct numerical answer with no work shown will receive only 1 credit. [15]

26 Solve for x: $x^2 + 3x - 40 = 0$

27 A person standing on level ground is 2,000 feet away from the foot of a 420-foot-tall building, as shown in the accompanying diagram. To the *nearest degree*, what is the value of x?

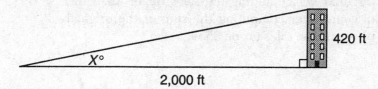

420 ft

$X°$

2,000 ft

28 Bob and Ray are describing the same number. Bob says, "The number is a positive even integer less than or equal to 20." Ray says, "The number is divisible by 4." If Bob's statement is true and Ray's statement is false, what are all the possible numbers?

29 Line ℓ contains the points (0,4) and (2,0). Show that the point (−25,81) does or does not lie on line ℓ.

30 A painting that regularly sells for a price of $55 is
on sale for 20% off. The sales tax on the painting is
7%. Will the final total cost of the painting differ
depending on whether the salesperson deducts the
discount before adding the sales tax or takes the
discount after computing the sum of the original
price and the sales tax on $55?

PART IV

**Answer all questions in this part. Each correct answer will
receive four (4) credits. Clearly indicate the necessary steps,
including appropriate formula substitutions, diagrams,
graphs, charts, etc. For all questions in this part, a correct
numerical answer with no work shown will receive only 1
credit.** [20]

31 The profits in a business are to be shared by the
three partners in the ratio of 3 to 2 to 5. The profit
for the year was $176,500. Determine the number
of dollars each partner is to receive.

32 If asphalt pavement costs $0.78 per square foot, determine, to the *nearest cent,* the cost of paving the shaded circular road with center O, an outside radius of 50 feet, and an inner radius of 36 feet, as shown in the accompanying diagram.

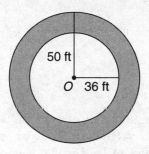

33 An arch is built so that it is 6 feet wide at the base. Its shape can be represented by a parabola with the equation $y = -2x^2 + 12x$, where y is the height of the arch.

a Graph the parabola from $x = 0$ to $x = 6$ on the grid below.

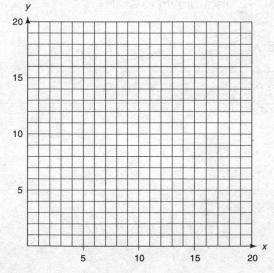

b Determine the maximum height, y, of the arch.

34 Mr. Gonzalez owns a triangular plot of land *BCD*
with *DB* = 25 yards and *BC* = 16 yards. He wishes
to purchase the adjacent plot of land in the shape
of right triangle *ABD*, as shown in the accompanying diagram, with *AD* = 15 yards. If the purchase is
made, what will be the total number of square
yards in the area of his plot of land, △*ACD*?

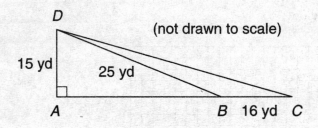

(not drawn to scale)

D

15 yd

25 yd

A

B　16 yd　C

35 Two health clubs offer different membership plans. The graph below represents the total cost of belonging to Club *A* and Club *B* for one year.

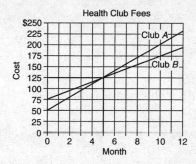

a If the yearly cost includes a membership fee plus a monthly charge, what is the membership fee for Club *A*?

b (1) What is the number of the month when the total cost is the same for both clubs?

(2) What is the total cost for Club *A* when both plans are the same?

c What is the monthly charge for Club *B*?

Answers
August 1999
Math A

Answer Key

PART I

1. (4)	**5.** (1)	**9.** (4)	**13.** (3)	**17.** (3)
2. (1)	**6.** (3)	**10.** (2)	**14.** (3)	**18.** (4)
3. (2)	**7.** (2)	**11.** (1)	**15.** (4)	**19.** (1)
4. (3)	**8.** (2)	**12.** (1)	**16.** (3)	**20.** (2)

PART II

21. 2.4

22. a. (B, B, B), (B, B, G), (B, G, B), (G, B, B), (C, C, C), (G, C, B), (G, B, G), and (B, G, G). For tree diagram, see Answers Explained section.

b. $\frac{1}{8}$

23. 72

24. 100

25. Two

PART III

26. −8 and 5

27. 12

28. 2, 6, 10, 14, and 18

29. Point (−25,81) is *not* on line ℓ.

30. No, the final cost will *not* differ.

PART IV

31. $52,950; $35,300; $88,250

32. $2,950.33

33. a. See Answers Explained
section.

    b. 18 feet

34. 270

35. a. $50 **b.** (1) 5 (2) $125
    c. $10

Parts II–IV You are required to show how you arrived at your answers. For sample methods of solution, see Answers Explained section.

Answers Explained

PART I

1. The basic shapes offered as answer choices—rectangular solid, pyramid, cone, and cylinder—are shown in the figures below.

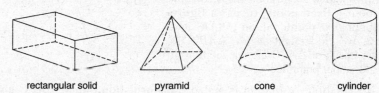

rectangular solid   pyramid   cone   cylinder

A roll of candy has two congruent circular bases (see the figure at the right) at opposite ends and is, therefore, a cylinder.

The correct choice is **(4)**.

2. The given expression is: $\sqrt{50}$

Factor the number underneath the radical as the product of two whole numbers, one of which is the largest perfect square factor of 50: $\sqrt{25 \cdot 2}$

Write the radical over each factor: $\sqrt{25} \cdot \sqrt{2}$

Simplify: $5\sqrt{2}$

The correct choice is **(1)**.

3. To determine the type of transformation shown in the accompanying diagram, consider each choice in turn:

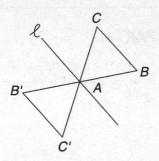

- (1): Line reflection in line ℓ. Since line ℓ is not the perpendicular bisector of $\overline{BB'}$ and $\overline{CC'}$, points B' and C' are not the images of points B and C, respectively, after a line reflection in line ℓ.

- (2): Rotation about point A. A rotation about a fixed point "turns" a figure about that point. Since $\angle BAB'$ and $\angle CAC'$ are straight angles, $\triangle AB'C'$ is the image of $\triangle ABC$ after a rotation of $180°$ about point A.

- (3): Dilation. A dilation is a size transformation. Since $\triangle ABC$ and $\triangle AB'C'$ appear to be the same size, the transformation is not a dilation.

- (4): Translation. Since $\triangle AB'C'$ cannot be obtained by "sliding" $\triangle ABC$ in the horizontal direction, the vertical direction, or both directions, the transformation is not a translation.

The correct choice is **(2)**.

4. Since each of the answer choices includes 10^{21} as a factor, rewrite the given expression, 6.02×10^{23}, so that it matches the form of the answer choices:

$$6.02 \times 10^{23} = 6.02 \times \overbrace{10^2 \times 10^{21}}^{10^{23}}$$
$$= 6.02 \times 100 \times 10^{21}$$
$$= 602 \times 10^{21}$$

The correct choice is **(3)**.

5. A regular pentagon has five sides with the same length. If the length of one side of the Pentagon building is represented by $n + 2$, then its perimeter is $5(n + 1)$ or, removing parentheses, $5n + 10$.

The correct choice is **(1)**.

6. To find the product of $4x^2y$ and $2xy^3$, multiply the numerical coefficients and then multiply like variable bases by adding their exponents:

$$(4x^2y)(2xy^3) = (4 \cdot 2)(x^2 \cdot x^1)(y^1 \cdot y^3)$$
$$= 8x^{2+1}y^{1+3}$$
$$= 8x^3y^4$$

The correct choice is **(3)**.

7. The additive identity property of real numbers states that there exists a number, namely 0, that, when added to any other number, call it x, produces a sum that is equal to the same number. x. In other words, $x + 0 = x$.

The correct choice is **(2)**.

8. The formula to convert from F degrees Fahrenheit to C degrees Celsius is given as $C = \dfrac{5}{9}(F - 32)$.

To find the Celsius temperature that is equal to a Fahrenheit temperature of 77°, substitute 77 for F in the given formula and solve for C:

$$\begin{aligned} C &= \frac{5}{9}(F - 32) \\ &= \frac{5}{9}(77 - 32) \\ &= \frac{5}{9}(45) \\ &= \frac{5}{\cancel{9}}(\cancel{45}^{5}) \\ &= 5 \times 5 \\ &= 25° \end{aligned}$$

The correct choice is **(2)**.

9. It is given that, in the accompanying diagram of rectangle $ABCD$, m$\angle BAC = 3x + 4$ and m$\angle ACD = x + 28$.

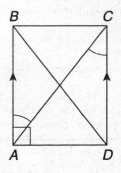

- The opposite sides of a rectangle are parallel. Since $AB \parallel CD$, the measures of alternate interior angles formed by transversal AC are equal. Thus:

$$\begin{aligned} m\angle BAC &= m\angle ACD \\ 3x + 4 &= x + 28 \\ 3x - x &= 28 - 4 \\ 2x &= 24 \\ x &= 12 \end{aligned}$$

- Because $x = 12$, m$\angle ACD = x + 28 = 12 + 28 = 40$.
- Since each of the four angles of a rectangle measures 90°, m$\angle BAD = 90$. Hence:

$$\begin{aligned} m\angle CAD + m\angle BAC &= m\angle BAD \\ m\angle CAD + m\angle BAC &= 90 \\ m\angle CAD + \quad 40 \quad &= 90 \\ m\angle CAD &= 90 - 40 = 50 \end{aligned}$$

The correct choice is **(4)**.

10. It is given that, on June 17, the temperature in New York City ranged from 90° to 99°, while the temperature in Niagara Falls ranged from 60° to 69°.

- The *greatest* possible difference in the temperatures of the two cities occurred when the temperature in New York City reached its highest and the temperature in Niagara Falls reached its lowest:

$$99° - 60° = 39°$$

- The *smallest* possible difference in the temperatures of the two cities occurred when the temperature in New York City reached its lowest and the temperature in Niagara Falls reached its highest:

$$90° - 69° = 21°$$

- At any given time on June 17, the difference in the temperatures in these two cities ranged from a low of 21° to a high of 39°. Thus, the difference in the temperatures in these two cities was *between* 20° and 40°.

The correct choice is (**2**).

11. The given sum is:

$$\frac{a}{x} + \frac{b}{2x}$$

The least common denominator (LCD) of the two fractions is $2x$. Change the first fraction into an equivalent fraction that has the LCD as its denominator by multiplying it by 1 in the form of $\frac{2}{2}$:

$$\frac{2}{2} \cdot \left(\frac{a}{x}\right) + \frac{b}{2x}$$

Simplify:

$$\frac{2a}{2x} + \frac{b}{2x}$$

Add fractions with the same denominator by writing the sum of their numerators over the common denominator:

$$\frac{2a+b}{2x}$$

The correct choice is (**1**).

12. A conditional statement and its converse may or may not have the same truth value.

- The given conditional statement, "If two angles are right angles, the angles have equal measure," is true since all right angles measure 90°.

- The converse of the given conditional statement, "If two angles have equal measure then the two angles are right angles," is false since two angles may have equal measure (say, 30° or 60°) that is different from 90°.

The correct choice is (**1**).

13. It is given that 6 and x have the same mean (average) as 2, 4, and 24.

- The average of 6 and x is the sum of the two numbers divided by 2, or $\frac{x+6}{2}$.

- The average of 2, 4, and 24 is the sum of the three numbers divided by 3, or $\frac{2+4+24}{3} = \frac{30}{3} = 10$.

- Thus:

$$\frac{x+6}{2} = 10$$

$$\cancel{2}\left(\frac{x+6}{\cancel{2}}\right) = 2(10)$$

$$x + 6 = 20$$

$$x = 20 - 6 = 14$$

The correct choice is **(3)**.

14. It is given that, in a hockey league, 87 players play on seven different teams, and that each team has at least 12 players.

- Since $12 \times 7 = 84$, there are $87 - 84 = 3$ players in the league above the required minimum number.

- If these three players are on the same team, that team will have $12 + 3 = 15$ players, which is, therefore, the largest possible number of players on any one team.

The correct choice is **(3)**.

15. It is given that, in the accompanying diagram of equilateral triangle ABC, $DE = 5$, $DE \parallel AB$, and AB is three times as long as DE; therefore, $AB = 3 \times 5 = 15$.

- Since $\triangle ABC$ is equilateral, and $AC = BC = AB = 15$ and m$\angle A$ = m$\angle B$ = m$\angle C$ = 60.

- Corresponding angles formed by parallel lines have equal measure, so m$\angle CDE$ = m$\angle A$ = 60 and m$\angle CED$ = m$\angle B$ = 60. Since $\triangle DCE$ is equiangular, it is also equilateral, so $CD = CE = DE = 5$.

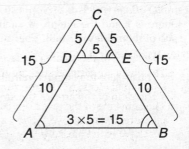

- Then $AD = AC - CD = 15 - 5 = 10$ and $EB = BC - CE = 15 - 5 = 10$.

• Therefore:

$$\text{Perimeter of quadrilateral } ABED = AD + DE + EB + AB$$
$$= 10 + 5 + 10 + 15$$
$$= 40$$

The correct choice is **(4)**.

16. It is given that at a concert twice as many hot dogs were sold as hamburgers, and three times as many soft drinks were sold as hamburgers.

Let x = the number of hamburgers sold.

Then $2x$ = the number of hot dogs sold,

and $3x$ = the number of soft drinks sold.

It is also given that all three items sold for $1.00 each and that the total amount of money collected for the hot dogs, hamburgers, and soft drinks was $720. Thus:

$$(x \cdot \$1) + (2x \cdot \$1) + (3x \cdot \$1) = \$720$$
$$\$x + \$2x + \$3x = \$720$$
$$\$6x = \$720$$
$$x = \frac{\$720}{\$6} = 120$$

Hence, the number of soft drinks sold = 3x = 3(120) = 360.

The correct choice is **(3)**.

17. Since a set of n different objects can be arranged in $n!$ ways, the number of different 6-letter arrangements that can be formed using the letters in the word "ABSENT," using each letter only once, is 6!, where

$$6! = 6 \times 5 \times 4 \times 3 \times 2 \times 1 = 720.$$

The correct choice is **(3)**.

18. It is given that the ratio of the lengths of corresponding sides of two similar squares is 1 to 3. Suppose that the length of each side of the smaller square is 1 and the length of each side of the larger square is 3, so that the ratio of their sides is 1 to 3. Then:

• The area of the smaller square is $1 \times 1 = 1$.

• The area of the larger square is $3 \times 3 = 9$.

• The ratio of the area of the smaller square to the area of the larger square is 1:9.

The correct choice is **(4)**.

19. If an equation of a line is in the form $y = mx + b$, then the slope of the line is m, the coefficient of the x-term, and the y-intercept of the line is b. To find the slope of the line whose equation is $3x - 4y - 16 = 0$, write the equation in the form $y = mx + b$ by solving for y:

The given equation is:
$$3x - 4y - 16 = 0$$

Add $4y$ to each side of the equation:
$$\frac{+ 4y \qquad = 0 + 4y}{3x \quad - 16 = \quad 4y}$$

or
$$4y = 3x - 16$$

Divide each member of the equation by 4:
$$\frac{4y}{4} = \frac{3x}{4} - \frac{16}{4}$$

$$y = \frac{3}{4}x - 4$$

For the equation $y = \frac{3}{4}x - 4$, $m = \frac{3}{4}$, so the slope of the given line is $\frac{3}{4}$.

The correct choice is **(1)**.

20. Since it is given that the height of an equilateral triangle is $2\sqrt{3}$, draw an equilateral triangle with a segment drawn from a vertex perpendicular to the opposite side, as shown in the accompanying diagram.

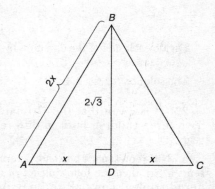

- In an equilateral triangle, the altitude drawn to the base bisects the base. If $AD = x$, then $AC = AB = BC = 2x$.

- In right triangle ADB, use the Pythagorean theorem:

$$(AB)^2 = (AD)^2 + (BD)^2$$
$$(2x)^2 = (x)^2 \quad + \left(2\sqrt{3}\right)^2$$
$$4x^2 = x^2 \quad + \left(2\sqrt{3}\right)\left(2\sqrt{3}\right)$$
$$4x^2 = x^2 \quad + 4 \cdot 3$$
$$4x^2 - x^2 = 12$$
$$3x^2 = 12$$
$$x^2 = \frac{12}{3} = 4$$
$$x = \sqrt{4} = 2$$

• Hence, $AC = AB = BC = 2(2) = 4$, so the perimeter of the equilateral triangle is $4 + 4 + 4 = 12$.

The correct choice is **(2)**.

PART II

21. The given equation is:

$$2(x - 3) = 1.2 - x$$

Remove the parentheses by multiplying each number inside the parentheses by the number in front of the parentheses:

$$2x - 6 = 1.2 - x$$

Collect like terms on the same side of the equation. First, add 6 to each side of the equation:

$$\underline{+ 6 = +6}$$

$$2x \quad = 7.2 - x$$

Next, add x to each side of the equation:

$$\underline{+x \quad = \quad +x}$$

$$3x \quad = 7.2$$

Divide each side of the equation by 3:

$$\frac{3x}{3} = \frac{7.2}{3}$$
$$x = 2.4$$

The solution for x is **2.4**.

22. It is given that the Grimaldis have three children born in different years.

a. If B represents a boy and G represents a girl, then the following set of ordered triples shows all the possible arrangements (the sample space) of boy and girl children in the Grimaldi family:

(B, B, B), (B, B, G), (B, G, B), (G, B, B), (G, G, G), (G, G, B), (G, B, G), (B, G, G)

b. Of the eight ordered triples, only one, (B, B, B), represents three boys.

Thus, the probability that the Grimaldis have three boys is $\frac{1}{8}$.

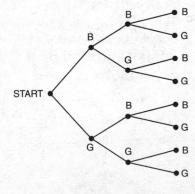

23. It is given that Paloma has 3 jackets, 6 scarves, and 4 hats.

- With each of the 3 jackets, any 1 of 6 scarves can be worn, so there are $3 \times 6 = 18$ jacket-scarf outfits.

- With each of the 18 jacket-scarf outfits, any 1 of 4 hats can be worn, so there are $18 \times 4 = 72$ jacket-scarf-hat outfits.

There are $3 \times 6 \times 4$ or **72** different outfits consisting of a jacket, a scarf, and a hat that Paloma can wear.

24. It is given that the accompanying circle graph represents the results of a poll in which 600 people were asked whether they like Chinese food.

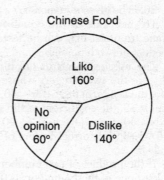

Chinese Food

- The measures of the central angles for the sectors that comprise a circle graph must add up to 360°. Hence, the measure of the central angle of the sector that represents "no opinion" is:

$360° - (160° + 140°) = 360° - 300° = 60$

- The sector for "no opinion" represents $\frac{60°}{360°} = \frac{1}{6}$ of the whole.

- Since $\frac{1}{6} \times 600 = 100$, the sector for "no opinion" represents 100 people.

In the poll **100** people had no opinion about Chinese food.

25. It is given that Maria's backyard has two trees that are 40 feet apart, and that Maria wants to place lampposts so that the posts are 30 feet from both of the trees.

- The locus of points 30 feet from tree A is a circle of radius 30 feet with tree A at its center, as shown in the accompanying diagram.

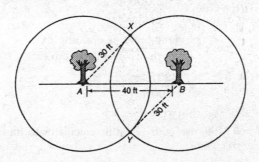

• The locus of points 30 feet from tree B is a circle of radius 30 feet with tree B at its center, as shown in the accompanying diagram.

• The two circles intersect at two different points, labeled X and Y.

Hence, **two** locations are possible where the lampposts could be placed in relation to the trees so that the lampposts are 30 feet from both trees.

PART III

26. The given equation is: $x^2 + 3x - 40 = 0$

The left side of the equation is a quadratic trinomial that can be factored as the product of two binomials: $(x + ?)(x + ?) = 0$

The missing terms of the binomial factors are the two integers whose product is −40, the last term of $x^2 + 3x - 40$, and whose sum is +3, the coefficient of the x-term of $x^2 + 3x - 40$. Since $(+8)(-5) = -40$ and $(+8) + (-5) = +3$, the correct factors of −40 are +8 and −5: $(x + 8)(x - 5) = 0$

If the product of two numbers is 0, then at least one of these numbers is 0: $x + 8 = 0$ or $x - 5 = 0$
$x = -8$ or $x = 5$

The two solutions for x are **−8** and **5**.

27. It is given that a person standing on level ground is 2,000 feet away from the foot of a 420-foot-tall building, as shown in the accompanying diagram.

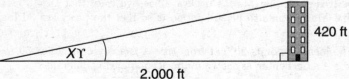

To find the value of x, use the tangent ratio:

$$\tan x° = \frac{\text{length of side opposite } \angle x}{\text{length of side adjacent to } \angle x}$$

$$= \frac{420 \text{ feet}}{2,000 \text{ feet}}$$

$$= 0.21$$

Since $\tan x° = 0.2100$, use your scientific calculator to find x. Depending on your particular calculator:

- Enter 0.21, press the key labeled $\boxed{\text{SHIFT}}$ or $\boxed{\text{INV}}$ or $\boxed{\text{2ND}}$, and then press $\boxed{\text{TAN}}$; or

- Press the key labeled $\boxed{\text{SHIFT}}$ or $\boxed{\text{INV}}$ or $\boxed{\text{2ND}}$, enter 0.21, press $\boxed{\text{TAN}}$, and then press $\boxed{=}$.

The display shows 11.859779. . . ; therefore, to the *nearest degree*, the value of x is **12**.

28. It is given that Bob describes a number by saying, "The number is a positive even integer less than or equal to 20," and Ray describes the same number by saying, "The number is divisible by 4."

- If Bob's statement is *true*, then the number may be one of the following:

$$2, 4, 6, 8, 10, 12, 14, 16, 18, 20$$

- If Ray's statement is *false*, then the number is *not* divisible by 4, so five numbers can be eliminated from Bob's list:

$$2, \cancel{4}, 6, \cancel{8}, 10, 1\cancel{2}, 14, 1\cancel{6}, 18, 2\cancel{0}$$

- The remaining numbers in Bob's list are 2, 6, 10, 14, and 18.

The possible numbers are **2, 6, 10, 14,** and **18**.

29. Three points lie on the same line if the slope of the line determined by the first and the second points is the same as the slope of the line determined by the second and the third points.

The slope of a line is the change in the y-coordinates of two points on the line divided by the change in the x-coordinates of the two points. Thus, the slope m of the line containing $A(x_A, y_A)$ and $B(x_B, y_B)$ is given by the formula

$$m = \frac{y_B - y_A}{x_B - x_A}$$

- It is given that line ℓ contains points (0,4) and (2,0). If $A(x_A, y_A) = (0,4)$ and $B(x_B, y_B) = (2,0)$, then:

$$\text{Slope of } \ell = \frac{\text{change in } y\text{-coordinates}}{\text{change in } x\text{-coordinates}}$$

$$= \frac{y_B - y_A}{x_B - x_A}$$

$$= \frac{0 - 4}{2 - 0}$$

$$= \frac{-4}{2}$$

$$= -2$$

• To find the slope of the line containing point (2,0) and the third point, (–25,81), let $A(x_A, y_A) = (2,0)$ and $B(x_B, y_B) = (-25,81)$. Then:

$$\text{Slope} = \frac{y_B - y_A}{x_B - x_A}$$

$$= \frac{81 - 0}{-25 - 2}$$

$$= -3$$

• Compare slopes. The slope of line ℓ is –2 . The slope of the line containing the second and third points, (2,0) and (–25,81), is –3.

Since the two slopes are not equal, point (–25,81) does **not** lie on line ℓ.

30. It is given that a painting that regularly sells for a price of $55 is on sale for 20% off, and the sales tax on the painting is 7%.

Case 1. Assume the discount is deducted before the sales tax is added.

• After a 20% discount, the price is $0.80 \times \$55 = \$44.00$.

• After salex tax is added, the final price of the painting is $44.00 + (0.07 \times \$44.00) = \$44.00 + \$3.08 = \$47.08$.

Case 2. Assume the sales tax is added before the discount is deducted.

• After the sales tax is added, the price is $55.00 + (0.07 \times \$55) = \$55.00 + \$3.85 = \$58.85$.

• After a 20% discount, the final price of the painting is $0.80 \times \$58.85 = \$47.08$.

• After the sales tax is added, the final price of the painting is $44.00 + (0.07 \times \$44.00) = \$44.00 + \$3.08 = \$47.08$.

In each case the final price of the painting is the same. Thus, the final total cost of the painting does **not** differ depending on whether the salesperson deducts the discount before adding the sales tax or takes the discount after computing the sum of the original price and the sales tax on $55.

PART IV

31. It is given that three partners in a business share the profits in the ratio of 3 to 2 to 5, and that the profit for the year was $176,500. Let $3x$, $2x$, and $5x$ represent the dollar amounts of the partners' shares of the yearly profit. Then

$$3x + 2x + 5x = \$176,500$$
$$10x = \$176,500$$
$$x = \frac{\$176,500}{10} = \$17,650$$

Thus:

$$3x = 3(\$17,650) = \$52,950$$
$$2x = 2(\$17,650) = \$35,300$$
$$5x = 5(\$17,650) = \$88,250$$

The partners receive **$52,950, $35,300,** and **$88,250**.

32. It is given that a circular road with center O, an outside radius of 50 feet, and an inner radius of 36 feet, as shown in the accompanying diagram, will be paved with asphalt that costs $0.78 per square foot.

The area A of a circle with radius r is given by the formula $A = \pi r^2$. Then:

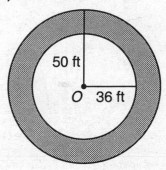

- The area of the circle with an outer radius of 50 feet is $\pi \times (50)^2 = 2,500\pi$ square feet.

- The area of the circle with an inner radius of 36 feet is $\pi \times (36)^2 = 1,296\pi$ square feet.

- The area of the shaded circular road is the difference between the areas of the outer and inner circles: $2,500\pi - 1,296\pi = 1,204\pi$ square feet.

Since the asphalt pavement costs $0.78 per square foot:

$$\text{Cost of paving shaded circular road} = \text{area of shaded circle} \times \$0.78$$
$$= 1204 \times \pi \times \$0.78$$
$$\approx \$2,950.33$$

The cost of paving the shaded circular road, correct to the *nearest cent,* is **$2,950.33**. [*Note:* Correct answers will vary depending on the approximation used for π.]

33. It is given that an arch is built so that it is 6 feet wide at the base and its shape can be represented by a parabola with the equation $y = -2x^2 + 12x$, where y represents the height of the arch.

a. To graph the parabola from $x = 0$ to $x = 6$, first prepare a table of values.

x	$-2x^2 + 12x =$	y
0	$-2(0)^2 + 12(0) =$	0
1	$-2(1)^2 + 12(1) = -2 + 12 =$	10
2	$-2(2)^2 + 12(2) = -2(4) + 24 = -8 + 24 =$	16
3	$-2(3)^2 + 12(3) = -2(9) + 36 = -18 + 36 =$	18
4	$-2(4)^2 + 12(4) = -2(16) + 48 = -32 + 48 =$	16
5	$-2(5)^2 + 12(5) = -2(25) + 60 = -50 + 60 =$	10
6	$-2(6)^2 + 12(6) = -2(36) + 72 = -72 + 72 =$	0

Using the graph paper provided, plot points (0,0), (1,10), (2,16), (3,18), (4,16), (5,10), and (6,0). Then connect these points with a smooth curve that has the shape of a parabola with its turning point at (3,18), as shown in the accompanying diagram.

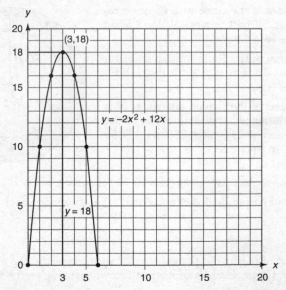

b. The maximum height, y, of the arch is at the turning point, (3,18), of the parabola where $y = \mathbf{18\ feet}$.

34. It is given that, in the accompanying diagram, BCD represents a triangular plot of land, $DB = 25$ yards, $BC = 16$ yards, and DAB represents an adjacent plot of land in the shape of a right triangle with $AD = 15$ yards.

• The lengths of the sides of right triangle DAB form a multiple of a 3 - 4 - 5 Pythagorean triple since:

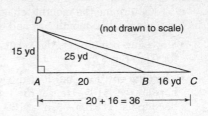

$$\text{leg } AD = 3 \times \underline{5} = 15$$

$$\text{leg } AB = 4 \times \underline{5} = 20$$

$$\text{hypotenuse } DB = 5 \times \underline{5} = 25$$

• Since $AB = 20$ yards, $AC = 20 + 16 = 36$ yards.

• Therefore:

$$\text{Area of right triangle } DAC = \frac{1}{2} \times AD \times AC$$

$$= \frac{1}{2} \times 15 \times 36$$

$$= 270 \text{ square yards}$$

The total number of square yards in the area of $\triangle ACD$ is **270**.

35.

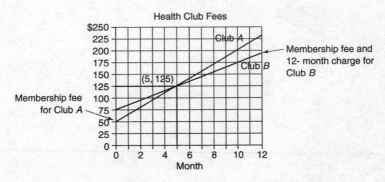

The total costs of belonging to Club A and Club B for one year are shown in the accompanying graph.

a. It is given that the yearly cost includes a membership fee plus a monthly charge. To figure out the membership fee for Club A, find the cost when the month number is 0. Since line A intersects the vertical (cost) axis at 50, the membership fee for Club A is **$50**.

b. (1) The total cost is the same for both clubs at the month where lines A and B intersect, that is, at the month with the number **5**.

(2) When the month number is 5, the cost for Club A is \$125. Hence, the total cost for Club A when both plans are the same is **\$125**.

c. To determine the monthly charge for Club B, subtract the membership fee from the total 12-month cost and then divide the result by 12.

- Since line B intersects the vertical (cost) axis at 75, the membership fee for Club B is \$75.

- The total cost for Club B at the end of month 12 is, including the membership fee, \$195. Hence, the total cost for the year, without the membership fee, is \$195 − \$75 = \$120.

The monthly charge for Club B is $\dfrac{\$120}{12} = \mathbf{\$10}$.

Topic	Question Numbers	Number of Points	Your Points	Your Percentage
1. Numbers; Properties of Real Numbers; Percent	7, 30	2 + 3		
2. Operations on Rat'l Numbers & Monomials	6	2		
3. Laws of Exponents for Integer Exponents; Scientific Notation	4	2		
4. Operations on Polynomials	—	—		
5. Square Root; Operations with Radicals	2	2		
6. Evaluating Formulas & Alg. Expressions	8	2		
7. Solving Linear Eqs. & Inequalities	21	2		
8. Solving Literal Eqs. & Formulas for a Particular Letter	—	—		
9. Alg. Operations (including factoring)	11	2		
10. Solving Quadratic Eqs.; Parabolas	26, 33	3 + 4		
11. Coordinate Geometry (Graphs of linear eqs.; slope; midpoint; distance)	19, 29	2 + 3		
12. Systems of Linear Eqs. & Inequalities (alg. and graph. solutions)	—	—		
13. Mathematical Modeling Using: Eqs., Tables, Graphs of Linear Eqs.	35	4		
14. Linear-Quad. Systems (alg. and graph. solutions)	—	—		
15. Word Problems Requiring Arith. or Alg. Reasoning	10, 14, 16, 28, 31	2 + 2 + 2 + 3 + 4		
16. Areas, Perims., Circums., Vols. of Common Figures	1, 5, 18, 20, 32. 34	2 + 2 + 2 + 2 + 4 + 4		
17. Angle & Line Relationships (suppl., compl., vertical angles; parallel lines; congruence)	—	—		
18. Ratio & Proportion (incl. similar polygons)	—	—		
19. Pythagorean Theorem	—	—		
20. Right Triangle Trig. & Indirect Measurement	27	3		
21. Logic (symbolic rep.; conditionals; logically equiv. statements; valid arguments)	12	2		
22. Probability (incl. tree diagrams & sample spaces)	22	2		
23. Counting Methods & Sets	23	2		
24. Permutations & Combinations	17	2		
25. Statistics (mean, percentiles, quartiles; freq. dist.; histograms; stem & leaf plots; circle graphs)	13, 24	2 + 2		
26. Properties of Triangles & Parallelograms	9, 15	2 + 2		
27. Transformations (reflections; translations; rotations; dilations)	3	2		

Topic	Question Numbers	Number of Points	Your Points	Your Percentage
28. Symmetry	—	—		
29. Area & Transformations Using Coordinates	—	—		
30. Locus & Constructions	25	2		
31. Dimensional Analysis	—	—		

HOW TO CONVERT YOUR RAW SCORE TO YOUR MATH A REGENTS EXAMINATION SCORE

Below is the conversion chart that must be used to determine your final score on the August 1999 Regents Examination in Mathematics A. To find your final exam score, locate in the column labeled "Raw Score" the total number of points you scored out of a possible 85 points. Since partial credit is allowed in Parts II, III, and IV of the test, you may need to approximate the credit you would receive for a solution that is not completely correct. Then locate in the adjacent column to the right the scaled score that corresponds to your raw score. The scaled score is your final Mathematics A Regents Examination score.

Regents Examination in Mathematics A—August 1999
Chart for Converting Total Test Raw Scores to
Final Examination Scores (Scaled Scores)

Raw Score	Scaled Score	Raw Score	Scaled Score	Raw Score	Scaled Score
85	100	56	74	27	45
84	99	55	73	26	44
83	99	54	71	25	42
82	98	53	71	24	42
81	98	52	70	23	41
80	97	51	69	22	40
79	97	50	68	21	39
78	96	49	67	20	38
77	95	48	66	19	36
76	94	47	65	18	37
75	93	46	64	17	35
74	92	45	63	16	34
73	91	44	62	15	33
72	90	43	61	14	32
71	89	42	60	13	31
70	88	41	59	12	30
69	87	40	58	11	29
68	86	39	57	10	28
67	85	38	56	9	27
66	84	37	55	8	24
65	83	36	54	7	21
64	82	35	53	6	18
63	81	34	52	5	15
62	80	33	51	4	12
61	79	32	50	3	9
60	78	31	49	2	6
59	77	30	48	1	3
58	76	29	47	0	0
57	75	28	46		

Examination
January 2000
Math A

PART I

Answer all questions in this part. Each correct answer will receive 2 credits. No partial credit will be allowed. Record your answers in the spaces provided. [40]

1 The expression $\sqrt{93}$ is a number between

(1) 3 and 9   (3) 9 and 10

(2) 8 and 9   (4) 46 and 47

1 _____

2 Which number has the greatest value?

(1) $1\frac{2}{3}$ (3) $\frac{\pi}{2}$

(2) $\sqrt{2}$ (4) 1.5

2 _____

3 Mary says, "The number I am thinking of is divisible by 2 or it is divisible by 3." Mary's statement is false if the number she is thinking of is

(1) 6   (3) 11

(2) 8   (4) 15

3 _____

4 Which expression is a factor of $x^2 + 2x - 15$?

(1) $(x - 3)$ (3) $(x + 15)$

(2) $(x + 3)$ (4) $(x - 5)$

4 _____

5 What was the median high temperature in Middletown during the 7-day period shown in the table below?

Daily High Temperature in Middletown	
Day	**Temperature (°F)**
Sunday	68
Monday	73
Tuesday	73
Wednesday	75
Thursday	69
Friday	67
Saturday	63

(1) 69   (3) 73
(2) 70   (4) 75   5 ____

6 If the number represented by $n - 3$ is an odd integer, which expression represents the next greater odd integer?

(1) $n - 5$ (3) $n - 1$
(2) $n - 2$ (4) $n + 1$ 6 ____

7 When the point $(2, -5)$ is reflected in the x-axis, what are the coordinates of its image?

(1) $(-5, 2)$ (3) $(2, 5)$
(2) $(-2, 5)$ (4) $(5, 2)$ 7 ____

8 The expression $(x^2 z^3)(x y^2 z)$ is equivalent to

(1) $x^2 y^2 z^3$ (3) $x^3 y^3 z^4$
(2) $x^3 y^2 z^4$ (4) $x^4 y^2 z^5$ 8 ____

9 Twenty-five percent of 88 is the same as what percent of 22?

(1) $12\frac{1}{2}\%$ (3) 50%

(2) 40%   (4) 100%   9 ____

10 A plot of land is in the shape of rhombus *ABCD* as shown below.

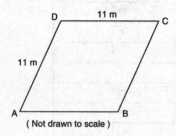

(Not drawn to scale)

Which can *not* be the length of diagonal *AC*?

(1) 24 m   (3) 11 m
(2) 18 m   (4) 4 m   10 ____

11 If $9x + 2a = 3a - 4x$, then x equals

(1) a (3) $\dfrac{5a}{12}$

(2) $-a$ (4) $\dfrac{a}{13}$ 11 ____

12 If the circumference of a circle is 10π inches, what is the area, in square inches, of the circle?

(1) 10π (3) 50π
(2) 25π (4) 100π 12 ____

13 How many different 4-letter arrangements can be formed using the letters of the word "JUMP," if each letter is used only once?

(1) 24   (3) 12
(2) 16   (4) 4   13 ____

14 Sterling silver is made of an alloy of silver and copper in the ratio of 37:3. If the mass of a sterling silver ingot is 600 grams, how much silver does it contain?

(1) 48.65 g   (3) 450 g
(2) 200 g   (4) 555 g   14 _____

15 If $t = -3$, then $3t^2 + 5t + 6$ equals

(1) −36   (3) 6
(2) −6   (4) 18   15 _____

16 The expression $\dfrac{y}{x} - \dfrac{1}{2}$ is equivalent to

(1) $\dfrac{2y - x}{2x}$ (3) $\dfrac{1 - y}{2x}$

(2) $\dfrac{x - 2y}{2x}$ (4) $\dfrac{y - 1}{x - 2}$ 16 _____

17 The party registration of the voters in Jonesville is shown in the table below.

Registered Voters in Jonesville	
Party Registration	**Number of Voters Registered**
Democrat	6,000
Republican	5,300
Independent	3,700

If one of the registered Jonesville voters is selected at random, what is the probability that the person selected is *not* a Democrat?

(1) 0.333   (3) 0.600
(2) 0.400   (4) 0.667   17 _____

18 If the number of molecules in 1 mole of a substance is 6.02×10^{23}, then the number of molecules in 100 moles is

(1) 6.02×10^{21} (3) 6.02×10^{24}
(2) 6.02×10^{22} (4) 6.02×10^{25} 18 ____

19 When $3a^2 - 2a + 5$ is subtracted from $a^2 + a - 1$, the result is

(1) $2a^2 - 3a + 6$ (3) $2a^2 - 3a - 6$
(2) $-2a^2 + 3a - 6$ (4) $-2a^2 + 3a + 6$ 19 ____

20 The distance between parallel lines ℓ and m is 12 units. Point A is on line ℓ. How many points are equidistant from lines ℓ and m and 8 units from point A?

(1) 1   (3) 3
(2) 2   (4) 4   20 ____

PART II

Answer all questions in this part. Each correct answer will receive 2 credits. Clearly indicate the necessary steps, including appropriate formula substitutions, diagrams, graphs, charts, etc. For all questions in this part, a correct numerical answer with no work shown will receive only 1 credit. [10]

21 The midpoint M of line segment AB has coordinates $(-3,4)$. If point A is the origin, $(0,0)$, what are the coordinates of point B? [The use of the accompanying grid is optional.]

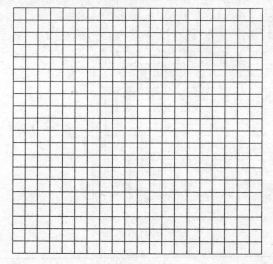

22 Mary and Amy had a total of 20 yards of material from which to make costumes. Mary used three times more material to make her costume than Amy used, and 2 yards of material was not used. How many yards of material did Amy use for her costume?

23 A wall is supported by a brace 10 feet long, as shown in the diagram below. If one end of the brace is placed 6 feet from the base of the wall, how many feet up the wall does the brace reach?

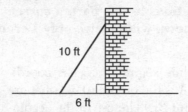

24 A straight line with slope 5 contains the points (1,2) and (3,K). Find the value of K. [The use of the accompanying grid is optional.]

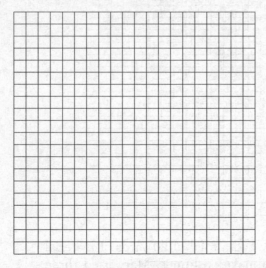

25 Al says, "If ABCD is a parallelogram, then ABCD is a rectangle." Sketch a quadrilateral ABCD that shows that Al's statement is *not* always true. Your sketch must show the length of each side and the measure of each angle for the quadrilateral you draw.

PART III

Answer all questions in this part. Each correct answer will receive 3 credits. Clearly indicate the necessary steps, including appropriate formula substitutions, diagrams, graphs, charts, etc. For all questions in this part, a correct numerical answer with no work shown will receive only 1 credit. [15]

26 Judy needs a mean (average) score of 86 on four tests to earn a midterm grade of B. If the mean of her scores for the first three tests was 83, what is the *lowest* score on a 100-point scale that she can receive on the fourth test to have a midterm grade of B?

27 A truck traveling at a constant rate of 45 miles per hour leaves Albany. One hour later a car traveling at a constant rate of 60 miles per hour also leaves Albany traveling in the same direction on the same highway. How long will it take for the car to catch up to the truck, if both vehicles continue in the same direction on the highway?

28 In the figure below, the large rectangle, $ABCD$, is divided into four smaller rectangles. The area of rectangle $AEHG = 5x$, the area of rectangle $GHFB = 2x^2$, the area of rectangle $HJCF = 6x$, segment $AG = 5$, and segment $AE = x$.

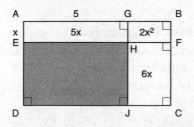

a Find the area of the shaded region.

b Write an expression for the area of rectangle *ABCD* in terms of *x*.

29 *a* On the set of axes provided below, sketch a circle with a radius of 3 and a center at (2,1) and also sketch the graph of the line $2x + y = 8$.

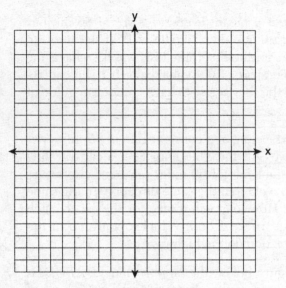

b What is the total number of points of intersection of the two graphs?

30 The volume of a rectangular pool is 1,080 cubic meters. Its length, width, and depth are in the ratio 10:4:1. Find the number of meters in each of the three dimensions of the pool.

PART IV

Answer all questions in the part. Each correct answer will receive 4 credits. Clearly indicate the necessary steps, including appropriate formula substitutions, diagrams, graphs, charts, etc. For all questions in this part, a correct numerical answer with no work shown will receive only 1 credit. [20]

31 Amy tossed a ball in the air in such a way that the path of the ball was modeled by the equation $y = -x^2 + 6x$. In the equation, y represents the height of the ball in feet and x is the time in seconds.

a Graph $y = -x^2 + 6x$ for $0 \le x \le 6$ on the grid provided below.

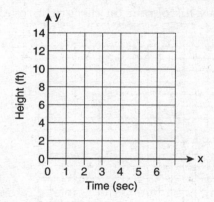

b At what time, x, is the ball at its highest point?

32 In the time trials for the 400-meter run at the state
sectionals, the 15 runners recorded the times shown
in the table below.

400-Meter Run	
Time (sec)	**Frequency**
50.0–50.9	
51.0–51.9	II
52.0–52.9	⊬III I
53.0–53.9	III
54.0–54.9	IIII

a Using the data from the frequency column, draw
a frequency histogram on the grid provided
below.

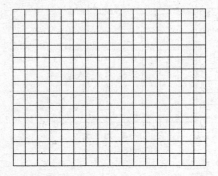

b What percent of the runners completed the time
trial between 52.0 and 53.9 seconds?

33 A group of 148 people is spending five days at a summer camp. The cook ordered 12 pounds of food for each adult and 9 pounds of food for each child. A total of 1,410 pounds of food was ordered.

   a Write an equation *or* a system of equations that describes the above situation and define your variables.

   b Using your work from part *a*, find:

   (1) the total number of adults in the group

   (2) the total number of children in the group

34 Three roses will be selected for a flower vase. The florist has 1 red rose, 1 white rose, 1 yellow rose, 1 orange rose, and 1 pink rose from which to choose.

   a How many different 3-rose selections can be formed from the 5 roses?

   b What is the probability that 3 roses selected at random will contain 1 red rose, 1 white rose, and 1 pink rose?

   c What is the probability that 3 roses selected at random will *not* contain an orange rose?

35 The Excel Cable Company has a monthly fee of $32.00 and an additional charge of $8.00 for each premium channel. The Best Cable Company has a monthly fee of $26.00 and an additional charge of $10.00 for each premium channel. The Horton family is deciding which of these two cable companies to subscribe to.

a For what number of premium channels will the total monthly subscription fee for the Excel and Best Cable companies be the same?

b The Horton family decides to subscribe to 2 premium channels for a period of one year.

(1) Which cable company should they subscribe to in order to spend less money?

(2) How much money will the Hortons save in one year by using the less expensive company?

Answers
January 2000
Math A

Answer Key

PART I

1. (3)	**5.** (1)	**9.** (4)	**13.** (1)	**17.** (3)
2. (1)	**6.** (3)	**10.** (1)	**14.** (4)	**18.** (4)
3. (3)	**7.** (3)	**11.** (4)	**15.** (4)	**19.** (2)
4. (1)	**8.** (2)	**12.** (2)	**16.** (1)	**20.** (2)

PART II

21. $(-6,8)$ or $x = -6, y = 8$

22. 4.5

23. 8

24. 12

25. See *Answers Explained* section.

PART III

26. 95 or 93

27. 3 hours

28. **a.** 15
   b. $(2x + 5)(x + 3)$ or
   $2x^2 + 11x + 15$

29. **a.** See *Answers Explained* section.
   b. 2

30. 3, 12, and 30

PART IV

31. a. See *Answers Explained* section.

   b. 3 seconds

32. a. See *Answers Explained* section.

   b. 60%

33. a. $12x + 9(148 - x) = 1,410$

   b. (1) 26   (2) 122

34. a. 10   **b.** $\dfrac{1}{10}$ or 0.1

   c. $\dfrac{4}{10}$ or $\dfrac{2}{5}$ or 0.4

35. a. 3

   b. (1) Best Cable Company

       (2) $24

Parts II–IV You are required to show how you arrived at your answers. For sample methods of solution, see Answers Explained section.

Answers Explained

PART I

1. The given expression, $\sqrt{\sqrt{93}}$, is the square root of $\sqrt{93}$. Use your calculator to find an approximation for $\sqrt{93}$. Since $\sqrt{93} \approx 9.64$, $\sqrt{93}$ is a number between 9 and 10.

The correct choice is (3).

2. To find which number among the four answer choices has the greatest value, express each number in decimal form:

- Choice (1): $1\dfrac{2}{3} \approx 1.667$

- Choice (2): $\sqrt{2} \approx 1.414$

- Choice (3): $\dfrac{\pi}{2} \approx 1.571$

- Choice (4): 1.5

Thus, $1\dfrac{2}{3}$ has the greatest value.

The correct choice is (1).

3. It is given that Mary says, "The number I am thinking of is divisible by 2 or it is divisible by 3." Mary's statement is false for any number that is not divisible by 2 or by 3. The number 11, given in choice (3), is not divisible by 2 and is also not divisible by 3.

The correct choice is (3).

4. The given expression is $x^2 + 2x - 15$.

- Factor $x^2 + 2x - 15$ as the product of two binomials in the form $(x + ?)$ $(x + ?)$. The missing numbers in the binomial factors are the two integers whose product is -15, the last term of $x^2 + 2x - 15$, and whose sum is $+2$, the coefficient of x in $x^2 + 2x - 15$.

- Since $(+5)(-3) = -15$ and $(+5) + (-3) = +2$, the two integers that satisfy these conditions are $+5$ and -3: $(x + 5)(x - 3)$.

- Examine the answer choices to see which one includes either $x + 5$ or $x - 3$. Choice (1) contains $(x - 3)$.

The correct choice is **(1)**.

5. The daily high temperatures in Middletown during a 7-day period are given in the accompanying table.

Daily High Temperature in Middletown	
Day	**Temperature** (°F)
Sunday	68
Monday	73
Tuesday	73
Wednesday	75
Thursday	69
Friday	67
Saturday	63

The median of a set of numbers is the middle value when the numbers are arranged in order from least to greatest.

- The 7 temperatures arranged from least to greatest are 63, 67, 68, <u>69</u>, 73, 73, 75.

- Since the median is the middle or fourth number in a list of 7 ordered numbers, the median temperature is 69.

The correct choice is **(1)**.

6. Since consecutive odd integers differ by 2, the next greater odd integer after $n - 3$ is $(n - 3) + 2 = n - 1$.

The correct choice is **(3)**.

7. When a point $P(x,y)$ is reflected in the x-axis, it is "flipped" over the x-axis so that the image of P is $P'(x,-y)$. That is, the x-coordinate of the point's image remains the same while the y-coordinate changes sign. Hence, when $(2,-5)$ is reflected in the x-axis, the coordinates of its image are $(2,5)$.

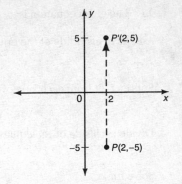

The correct choice is **(3)**.

8. The given expression is:

$$(x^2z^3)(xy^2z)$$

Rewrite the expression so that like factors are grouped together:

$$(x^2z^3)(xy^2z) = (x^2x^1)(y^2)(z^3z^1)$$

To multiply powers of the same base, keep the base and add their exponents:

$$(x^2z^3)(xy^2z) = x^3y^2z^4$$

The correct choice is **(2)**.

9. Twenty-five percent of 88 is $0.25 \times 88 = 22$. One hundred percent of 22 is also 22. Hence, 25% of 88 is the same as 100% of 22.

The correct choice is **(4)**.

10. It is given that a plot of land is in the shape of rhombus $ABCD$, as shown in the accompanying diagram.

Since the length of each side of a triangle must be less than the sum of the lengths of the other two sides, the length of the diagonal AC must be less than 11 m $+ 11$ m $= 22$ m. Hence, AC cannot be 24 m.

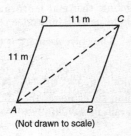

(Not drawn to scale)

The correct choice is **(1)**.

11. The given equation is: $9x + 2a = 3a - 4x$

Add $4x$ to each side of the equation:

$$+ 4x \quad\quad = + 4x$$
$$13x + 2a = 3a$$

Subtract $2a$ from each side of the equation:

$$- 2a = - 2a$$
$$13x \quad\quad = a$$

Divide each side of the equation by 13:

$$\frac{13x}{13} = \frac{a}{13}$$

Solve for x:

$$x = \frac{a}{13}$$

The correct choice is **(4)**.

12. It is given that the circumference of a circle is 10π inches.

• Since circumference is equal to π times diameter, the diameter of the circle is 10 inches. Hence, the radius of the circle is $\frac{1}{2} \times 10 = 5$ inches.

• The area of a circle is equal to $\pi \times (\text{radius})^2$. Since the radius is 5 inches, the area of the circle is $\pi \times 5^2 = 25\pi$ square inches.

The correct choice is **(2)**.

13. In general, n different objects can be arranged in $n!$ ways when each object is used only once. Hence, the number of different 4-letter arrangements that can be formed from the letters of the word JUMP when each letter is used only once is $4! = 4 \times 3 \times 2 \times 1 = 24$.

The correct choice is **(1)**.

14. It is given that sterling silver is made of an alloy of silver and copper in the ratio 37:3. If x represents the amount of silver in a 600 gram mass of sterling silver ingot, then $600 - x$ represents the amount of copper in the 600 gram mass. Thus:

$$\frac{\text{silver}}{\text{copper}} = \frac{x}{600 - x} = \frac{37}{3}$$

• In a proportion, the product of the means is equal to the product of the extremes. Cross-multiply and solve for x:

$$3x = 37(600 - x)$$
$$3x = 22,200 - 37x$$
$$3x + 37x = 22,200$$
$$40x = 22,200$$
$$x = \frac{22,200}{40} = 555$$

The correct choice is **(4)**.

15. To evaluate $3t^2 + 5t + 6$ when $t = -3$, replace t with -3 and simplify:

$$3(-3)^2 + 5(-3) + 6 = 3(9) - 15 + 6$$
$$= 27 - 15 + 6$$
$$= 12 + 6$$
$$= 18$$

The correct choice is **(4)**.

16. The given expression is:

$$\frac{y}{x} - \frac{1}{2}$$

To combine fractions, change each fraction into an equivalent fraction that has the least common denominator (LCD) as its denominator. The LCD of x and 2 is $2x$, since $2x$ is the smallest expression into which x and 2 both divide evenly.

Multiply the first fraction in the given expression by 1 in the form of $\frac{2}{2}$, and multiply the second fraction by 1 in the form of $\frac{x}{x}$:

$$\frac{2}{2}\left(\frac{y}{x}\right) - \frac{1}{2}\left(\frac{x}{x}\right)$$

Simplify:

$$\frac{2y}{2x} - \frac{x}{2x}$$

Write the difference of the numerators over the common denominator:

$$\frac{2y - x}{2x}$$

The correct choice is **(1)**.

17. The table shows the party registration of the voters in Jonesville.

Registered Voters in Jonesville	
Party	Number of Voters Registered
Democrat	6,000
Republican	5,300
Independent	3,700

According to the given table:

- The total number of voters registered is $6,000 + 5,300 + 3,700 = 15,000$.

- The total number of voters who are registered as Republican or Independent is $5,300 + 3,700 = 9,000$.

- Since 9,000 of the 15,000 voters are *not* registered as Democrats, the probability that a registered voter selected at random will *not* be a Democrat is $\dfrac{9,000}{15,000} = 0.600$.

The correct choice is **(3)**.

18. It is given that the number of molecules in 1 mole of a substance is 6.02×10^{23}. The number of molecules in 100 moles of the same substance is, therefore,

$$100 \times (6.02 \times 10^{23}) = 10^2 \times 6.02 \times 10^{23}$$

To multiply powers of the same base, keep the base and add the exponents:

$$6.02 \times 10^2 \times 10^{23} = 6.02 \times 10^{25}$$

The correct choice is **(4)**.

19. To subtract $3a^2 - 2a + 5$ from $a^2 + a - 1$, write the first polynomial underneath the second, aligning like terms in the same column. Change to an equivalent addition example by writing the opposite of each term of the second polynomial. Then add like terms in each column:

$$
\begin{array}{c}
a^2 + a - 1 \\
\underline{3a^2 - 2a + 5}
\end{array}
\quad
\xrightarrow[\text{addition example}]{\text{change to an}}
\quad
\begin{array}{c}
a^2 + a - 1 \\
\underline{-3a^2 + 2a - 5} \\
-2a^2 + 3a - 6
\end{array}
$$

The correct choice is **(2)**.

20. It is given that the distance between parallel lines ℓ and m is 12 units, and that point A is on line ℓ. To find the number of points that are equidistant from lines ℓ and m, and 8 units from point A, represent each locus condition on the same diagram. Then count the number of points of intersection of the two locus conditions.

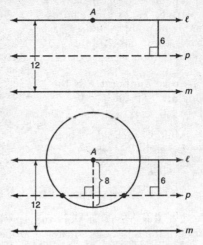

- Sketch $\ell \parallel m$ 12 units apart. Pick a point on line ℓ and label it A.

- The locus of points equidistant from lines ℓ and m is a parallel line midway between them. Label this line p, as shown in the upper diagram.

- The locus of points 8 units from point A is a circle with A as its center and a radius of 8 units.

- Since the radius of the circle is greater than the distance from line p to line ℓ, as shown in the lower diagram, the circle intersects line p at two different points.

- Since the two locus conditions intersect in two points, there are two points equidistant from lines ℓ and m and 8 units from point A.

The correct choice is **(2)**.

PART II

21. The x- and y-coordinates of the midpoint of a line segment are the averages of the corresponding coordinates of the endpoints of the line segment. It is given that the midpoint M of line segment AB has coordinates $(-3,4)$ and that point A is the origin, $(0,0)$.

If (x_A, y_A) are the coordinates of point A and (x_B, y_B) are the coordinates of point B, then:

$$-3 = \frac{x_A + x_B}{2} \qquad \text{and} \qquad 4 = \frac{y_A + y_B}{2}$$

Since endpoint A is the origin, $x_A = 0$ and $y_A = 0$:

$$-3 = \frac{0 + x_B}{2} \qquad\qquad\qquad 4 = \frac{0 + y_B}{2}$$

Multiply each side of both equations by 2:

$$-3 \times 2 = 0 + x_B \qquad\qquad 4 \times 2 = 0 + y_B$$

Simplify:

$$-6 = x_B \qquad\qquad\qquad 8 = y_B$$

The coordinates of point B are **(–6,8)** or $x = -6, y = 8$.

22. It is given that Mary and Amy had a total of 20 yards of material. Mary used three times more material to make her costume than Amy used, and 2 yards of material were not used.

If x represents the amount of material Amy used, then $3x$ represents the amount of material Mary used. Since 2 yards of material were not used, a total of $20 - 2 = 18$ yards were used to make both costumes. Hence:

$$x + 3x = 18$$
$$4x = 18$$
$$x = \frac{18}{4} = 4.5$$

Amy uses **4.5** yards of material.

23. It is given that a wall is supported by a brace 10 feet long and that one end of the brace is placed 6 feet from the base of the wall. Thus, the brace is the hypotenuse of a right triangle, as shown in the accompanying diagram. One side of the triangle is the distance of the brace from the base of the wall (6 ft) and the other side is the number of feet up the wall that the brace reaches (x).

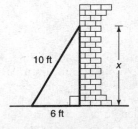

Method 1: Use a Pythagorean triple

The brace forms the hypotenuse of a right triangle in which the length of each of the three sides is two times as great as the corresponding dimension in a 3-4-5 Pythagorean triple:

- The hypotenuse is $2 \times \underline{5} = 10$ feet.
- One side is $2 \times \underline{3} = 6$ feet.
- The remaining side, x, must be $2 \times \underline{4} = 8$ feet.

The brace reaches **8** feet up the wall.

Method 2: Use the Pythagorean theorem

If x represents the number of feet up the wall the brace reaches, then x is one leg of a right triangle with another leg = 6 and hypotenuse = 10. By the Pythagorean theorem:

$$x^2 + y^2 = z^2$$

Substitute known values and solve for x:

$$x^2 + 6^2 = 10^2$$
$$x^2 + 36 = 100$$
$$x^2 = 100 - 36$$
$$x = \sqrt{64} = 8$$

The brace reaches **8** feet up the wall.

24. It is given that a straight line with slope 5 contains the points $(1,2)$ and $(3,K)$. The slope of a line is equal to the difference in the y-coordinates of two points divided by the difference in the x-coordinates of the two points. Thus:

$$\frac{\text{difference in } y\text{-coordinates}}{\text{difference in } x\text{-coordinates}} = \frac{K-2}{3-1} = 5$$
$$\frac{K-2}{2} = 5$$
$$K - 2 = 5 \times 2$$
$$K - 2 = 10$$
$$K = 10 + 2 = 12$$

The value of K is **12**.

25. It is given that Al states, "If $ABCD$ is a parallelogram, then $ABCD$ is a rectangle." To show that Al's statement is not always true, sketch a quadrilateral that has opposite sides of equal length and opposite angles of equal measure, but that does not contain four right angles. One such quadrilateral, a parallelogram but not a rectangle, is shown in the accompanying diagram.

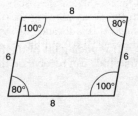

PART III

26. It is given that Judy needs a mean (average) score of 86 on four tests to earn a midterm grade of B, and that the mean of her scores for the first three tests was 83.

Let x represent the score Judy needs on the fourth test to have a midterm grade of B. The mean of a set of scores is the sum of the scores divided by the number of scores. Since the mean of Judy's first three test scores was 83, the sum of the three scores was $3 \times 83 = 249$.

Method 1: Assume no rounding off of the final test average is allowed

The lowest score needed to receive a B, assuming no rounding off of the average, is the score that makes the average of the four test scores 86. Thus:

$$\frac{\text{sum of four test scores}}{4} = 86$$

$$\frac{x + 249}{4} = 86$$

$$x + 249 = 86 \times 4$$

$$x + 249 = 344$$

$$x = 344 - 249$$

$$x = 95$$

The lowest score that Judy can receive on the fourth test to have a midterm grade of B is **95**.

Method 2: Assume rounding off of the final test average is allowed

The lowest score needed to receive a B is the score that makes the average of the four test scores 85.5, since 85.5 becomes 86 after rounding off. Thus:

$$\frac{\text{sum of four test scores}}{4} = 85.5$$

$$\frac{x + 249}{4} = 85.5$$

$$x + 249 = 85.5 \times 4$$

$$x + 249 = 342$$

$$x = 342 - 249$$

$$x = 93$$

If rounding off of the final test average is permitted, then the lowest score that Judy can receive on the fourth test to have a midterm grade of B is **93**.

27. It is given that a truck traveling at a constant rate of 45 miles per hour leaves Albany and that one hour later a car traveling at a constant rate of 60 miles per hour also leaves Albany, traveling in the same direction on the same highway.

If x represents the number of hours it takes the car to catch up to the truck, then $x + 1$ represents the number of hours the truck has traveled when it is overtaken by the car. The distances traveled are shown in the table.

	rate	× time	= distance
Truck	45 mph	$x + 1$	$45(x + 1)$
Car	60 mph	x	$60x$

Since the distances traveled by the truck and the car must be equal when they meet:

$$60x = 45(x + 1)$$
$$60x = 45x + 45$$
$$60x - 45x = 45$$
$$15x = 45$$
$$x = \frac{45}{15} = 3$$

The car will catch up to the truck **3 hours** after it leaves Albany.

28. It is given that, in the accompanying diagram of rectangle $ABCD$, the area of rectangle $AEHG = 5x$, the area of rectangle $GHFB = 2x^2$, the area of rectangle $HJCF = 6x$, segment $AG = 5$, and segment $AE = x$.

a. The area of the shaded region, rectangle $EDJH$, is the product of its length, EH, and its width, HJ. Use the information from adjacent rectangles to find these dimensions.

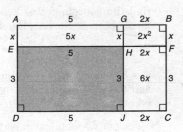

- Since $AEHG$ is a rectangle, **$EH =$** $AG = 5$ and $GH = AE = x$.
- The area of rectangle $GHFB$ is $2x^2$ and $GH = x$, so $HF = GB = 2x$.
- Since the area of rectangle $HJCF = 6x$, **$HJ = ED = 3$.**
- Hence, the area of rectangle $EDJH = EH \times HJ = 5 \times 3 = 15$ square units.

The area of the shaded region is **15** square units.

b. The area of rectangle $ABCD$ is the product of its length, AB, and its width, AD. Since $AB = GB + AG$ and $AD = AE + ED$:

$$\text{area of rectangle } ABCD = (AB)(AD) = (GB + AG)(AE + ED)$$

From part (**a**) you know that $GB = 2x$ and $ED = 3$. Thus:

$$\text{area of rectangle } ABCD = (GB + AG)(AE + ED)$$
$$= (2x + 5)(x + 3)$$

The area of rectangle $ABCD$, in terms of x, is $(2x + 5)(x + 3)$ or $2x^2 + 11x + 15$.

29. The given figures are a circle with a radius of 3 and a center at $(2,1)$, and the line $2x + y = 8$.

a. Sketch the circle and the line using the set of axes provided.

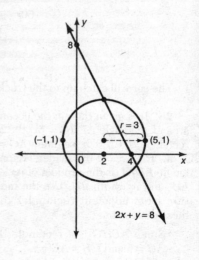

- To sketch a circle with a radius of 3 and a center at $(2,1)$, place the pivot point of your compass at $(2,1)$. Then set the radius length to 3 so that the pencil point reaches the point $(5,1)$, which is 3 units horizontally to the right of $(2,1)$. Keeping the pivot point of the compass fixed at $(2,1)$, draw a circle with the radius length fixed to 3 units, as shown in the accompanying diagram.

- To sketch the graph of $2x + y = 8$, find the coordinates of two convenient points on the line. If $y = 0$, then $2x = 8$, so $x = \dfrac{8}{2} = 4$. If $x = 0$, then $y = 8$. Hence, plot $(4,0)$ and $(0,8)$, and then draw a line through these points, as shown in the diagram.

b. The graphs sketched in part (**a**) intersect at **2** points.

30. It is given that the volume of a rectangular pool is 1,080 cubic meters, and that its length, width, and depth are in the ratio 10:4:1.

- Let $10x$ = the length of the pool in meters.
- Let $4x$ = the width of the pool in meters.
- Let x = the depth of the pool in meters.

Since the volume of a rectangular pool is the product of its length, width, and depth:

$$(10x)(4x)(x) = 1,080$$
$$40x^3 = 1,080$$
$$x^3 = \frac{1,080}{40}$$
$$x^3 = 27$$
$$x = \sqrt[3]{27} = 3$$

Since $x = 3$, $4x = 4(3) = 12$ and $10x = 10(3) = 30$.

The pool is **3** m deep, **12** m wide, and **30** m long.

PART IV

31. It is given that Amy tossed a ball in the air in such a way that the path of the ball was modeled by the equation $y = -x^2 + 6x$, where y represents the height of the ball in feet and x is the time in seconds.

a. To graph the equation $y = -x^2 + 6x$ for $0 \; " \; x \; " \; 6$, first prepare a table of integer values from $x = 0$ to $x = 6$.

x	$-x^2 + 6x =$	y
0	$-(0)^2 + 6(0) = 0 + 0 =$	0
1	$-(1)^2 + 6(1) = -(1) + 6 =$	5
2	$-(2)^2 + 6(2) = -(4) + 12 =$	8
3	$-(3)^2 + 6(3) = -(9) + 18 =$	9
4	$-(4)^2 + 6(4) = -(16) + 24 =$	8
5	$-(5)^2 + 6(5) = -(25) + 30 =$	5
6	$-(6)^2 + 6(6) = -(36) + 36 =$	0

Using the grid provided, plot the points (0,0), (1,5), (2,8), (3,9), (4,8), (5,5) and (6,0). Then connect these points with a smooth curve that has the shape of a parabola with its turning point at (3,9), as shown in the accompanying diagram.

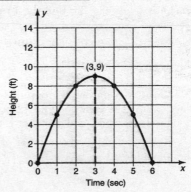

b. The ball will reach its highest point at the turning point, (3,9), of the parabola when $x = 3$ **seconds**.

32. It is given that at the time trials for the 400-meter run at the state sectionals, the 15 runners recorded the times shown in the table below.

400-Meter Run	
Time (sec)	**Frequency**
50.0–50.9	
51.0–51.9	||
52.0–52.9	＋Ht |
53.0–53.9	|||
54.0–54.9	||||

a. Using the data in the table, draw a frequency histogram on the grid provided, as shown in the accompanying diagram.

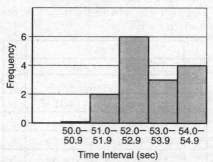

b. According to the given table, 6 runners had times in the interval from 52.0 to 52.9 seconds and 3 runners had times in the interval from 53.0 to 53.9 seconds. Thus, $6 + 3 = 9$ of the 15 runners had times between 52.0 and 53.9 seconds.

Since $\frac{9}{15} = 0.60 = 60\%$, **60%** of the runners completed the time trial between 52.0 and 53.9 seconds.

33. It is given that a group of 148 people is spending five days at a summer camp. The cook ordered 12 pounds of food for each adult, 9 pounds of food for each child, and a total of 1,410 pounds of food.

a. If x represents the number of adults, then $148 - x$ represents the number of children.

$$\underbrace{12x}_{\substack{\text{Pounds of food}\\\text{ordered for adults}}} + \underbrace{9(148 - x)}_{\substack{\text{Pounds of food}\\\text{ordered for children}}} = \underbrace{1,410}_{\substack{\text{Total pounds of}\\\text{food ordered}}}$$

The situation is described by the equation $12x + 9(148 - x) = 1,410$, where $x =$ number of adults and $148 - x =$ number of children.

b. **(1)** To find the total number of adults in the group, solve the equation written in part (**a**) for x:

$$12x + 9(148 - x) = 1,410$$
$$12x + 1,332 - 9x = 1,410$$
$$3x + 1,332 = 1,410$$
$$\underline{-1,332 = 1,332}$$
$$3x \qquad = 78$$

$$x = \frac{78}{3} = 26$$

The total number of adults in the group is **26**.

(2) Since $148 - x = 148 - 26 = 122$, the total number of children in the group is **122**.

34. It is given that a florist has 5 roses—1 red rose, 1 white rose, 1 yellow rose, 1 orange rose, and 1 pink rose—from which to choose 3 roses for a flower vase.

Method 1: Use combinations

a. Without considering order, r objects can be selected from a set of n objects $(r \leq n)$ in $_nC_r$ ways, where $_nC_r$ is the combination of n things taken r at a time. $_nC_r$ can be determined by calculator or by the formula $_nC_r = \dfrac{n!}{r!(n-r)!}$.

To find the number of different 3-rose selections that can be formed from the 5 given roses, use your calculator to evaluate $_nC_r$ when $n = 5$ and $r = 3$, thereby obtaining $_5C_3 = 10$.

The number of 3-rose selections that can be formed from the 5 roses is **10**.

b. Since the set of 5 roses contains only 1 red rose, only 1 white rose, and only 1 pink rose, only 1 of the 10 possible selections of 3 roses will contain 1 red rose, 1 white rose, and 1 pink rose.

The probability that 3 roses selected at random will contain 1 red rose, 1 white rose, and 1 pink rose is $\dfrac{1}{10}$ or **0.1**.

c. There are 4 roses that are not orange. The number of 3-rose selections that do *not* include an orange rose is, therefore, $_4C_3 = 4$. Hence, 4 of the 10 possible selections of 3 roses will *not* contain an orange rose.

The probability that 3 roses selected at random will *not* contain an orange rose is $\dfrac{4}{10}$ or $\dfrac{2}{5}$ or **0.4**.

Method 2: List the sample space

a. To determine the number of 3-rose selections that can be formed from the 5 roses, list all the possible 3-color combinations of roses in which the order of the colors is not considered:

1. (red, white, yellow)
2. (red, white, orange)
3. (red, white, pink)
4. (white, yellow, orange)
5. (white, yellow, pink)
6. (yellow, red, pink)
7. (yellow, orange, pink)
8. (orange, yellow, pink)
9. (orange, white, pink)
10. (pink, orange, red)

Hence, the number of different selections that can be formed from the 5 roses is **10**.

b. Only 1 of the 10 selections contains 1 red rose, 1 white rose, and 1 pink rose. Hence, the probability that 3 roses selected at random will contain 1 red rose, 1 white rose, and 1 pink rose is $\dfrac{1}{10}$ or **0.1**.

c. According to the sample space, 4 of the 10 selections do *not* include an orange rose. Hence, the probability that 3 roses selected at random will *not* contain an orange rose is $\dfrac{4}{10}$ or $\dfrac{2}{5}$ or **0.4**.

35. It is given that the Excel Cable Company has a monthly fee of $32.00 and an additional charge of $8.00 for each premium channel. The Best Cable Company has a monthly fee of $26.00 and an additional charge of $10.00 for each premium channel.

a. If x represents the number of premium channels a family orders, then:

- $32 + 8x$ represents the total monthly subscription fee for the Excel Cable Company.

- $26 + 10x$ represents the total monthly subscription fee for the Best Cable Company.

- The equation $26 + 10x = 32 + 8x$ represents the condition that the two cable companies will have the same total monthly subscription fees for x premium channels.

Solve for x:

$$26 + 10x = 32 + 8x$$
$$\underline{-26 \qquad\quad = -26}$$
$$10x = 6 + 8x$$
$$\underline{-8x = \qquad -8x}$$
$$2x = 6$$
$$x = \frac{6}{2} = 3$$

The total monthly subscription fees will be the same for **3** premium channels.

b. It is given that the Horton family decides to subscribe to 2 premium channels for a period of one year.

(1) To find the total monthly charges from each cable company for 2 premium channels, substitute 2 for x in $32 + 8x$ and $26 + 10x$ found in part **(a)**:

• The total monthly charges using the Excel Cable Company are

$$32 + 8x = \$32 + \$8 \times 2 = \$32 + \$16 = \$48$$

• The total monthly charges using the Best Cable Company are

$$26 + 10x = \$26 + \$10 \times 2 = \$26 + \$20 = \$46$$

• Hence, the Horton family will spend less money by subscribing to the **Best Cable Company**.

(2) The Best Cable Company charges $48 – \$46 = \$2$ less per month than the Excel Cable Company. Since there are 12 months in one year, the Hortons will save $12 \times \$2 = \textbf{\$24}$ in one year using the less expensive company.

Topic	Question Numbers	Number of Points	Your Points	Your Percentage
1. Numbers; Properties of Real Numbers; Percent	2, 6, 9	2 + 2 + 2 = 6		
2. Operations on Rat'l Numbers & Monomials	8, 16	2 + 2 = 4		
3. Laws of Exponents for Integer Exponents; Scientific Notation	18	2		
4. Operations on Polynomials	19	2		
5. Square Root; Operations with Radicals	1	2		
6. Evaluating Formulas & Alg. Expressions	15	2		
7. Solving Linear Eqs. & Inequalities	—	—		
8. Solving Literal Eqs. & Formulas for a Particular Letter	11	2		
9. Alg. Operations (including factoring)	4	2		
10. Solving Quadratic Eqs.	—	—		
11. Coordinate Geometry (graphs of linear eqs.; slope; midpoint; distance)	21, 24	2 + 2 = 4		
12. Systems of Linear Eqs. & Inequalities (alg. and graph. solutions)	—	—		
13. Mathematical Modeling Using: Eqs., Tables, Graphs of Linear Eqs., Parabolas	31, 35	4 + 4 = 8		
14. Linear-Quad Systems (alg. and graphical solutions)	29	3		
15. Word Problems Requiring Arith. or Algebraic Reasoning	22, 27, 33	2 + 3 + 4 = 9		
16. Areas, Perims., Circumf., Vols. of Common Figures	12, 28	2 + 3 = 5		
17. Angle & Line Relationships (suppl., compl., vertical angles; parallel lines; congruence)	—	—		
18. Ratio & Proportion (incl. similar polygons)	14, 30	2 + 3 = 5		
19. Pythagorean Theorem	23	2		
20. Right Triangle Trig. & Indirect Measurement	—	—		
21. Logic (symbolic rep.; conditionals; logically equiv. statements; valid arguments)	3	2		
22. Probability (incl. tree diagrams & sample spaces)	17	2		
23. Counting Methods & Sets	—	—		
24. Permutations & Combinations	13, 34	2 + 4 = 6		
25. Statistics (mean, percentiles, quartiles; freq. dist.; histograms; stem & leaf plots)	5, 26, 32	2 + 3 + 4 = 9		
26. Properties of Triangles & Parallelograms	10, 25	2 + 2 = 4		
27. Transformations (reflections; translations; rotations; dilations)	7	2		

	Topic	Question Numbers	Number of Points	Your Points	Your Percentage
28.	Symmetry	—	—		
29.	Area & Transformations Using Coordinates	—	—		
30.	Locus & Constructions	20	2		
31.	Dimensional Analysis	—	—		

HOW TO CONVERT YOUR RAW SCORE TO YOUR MATH A REGENTS EXAMINATION SCORE

Below is the conversion chart that must be used to determine your final score on the January 2000 Regents Examination in Math A. To find your final exam score, locate in the column labeled "Raw Score" the total number of points you scored out of a possible 85 points. Since partial credit is allowed in Parts II, III, and IV of the test, you may need to approximate the credit you would receive for a solution that is not completely correct. Then locate in the adjacent column to the right the scaled score that corresponds to your raw score. The scaled score is your final Math A Regents Examination score.

Regents Examination in Math A—January 2000
Chart for Converting Total Test Raw Scores to
Final Examination Scores (Scaled Scores)

Raw Score	Scaled Score	Raw Score	Scaled Score	Raw Score	Scaled Score
85	100	56	77	27	47
84	100	55	76	26	46
83	99	54	75	25	45
82	99	53	74	24	44
81	98	52	73	23	43
80	98	51	72	22	42
79	97	50	71	21	41
78	96	49	70	20	40
77	96	48	69	19	39
76	95	47	68	18	38
75	94	46	67	17	37
74	93	45	66	16	36
73	92	44	65	15	35
72	91	43	64	14	34
71	91	42	63	13	33
70	90	41	62	12	32
69	89	40	61	11	31
68	88	39	60	10	29
67	87	38	59	9	28
66	86	37	58	8	27
65	86	36	57	7	26
64	85	35	56	6	25
63	84	34	55	5	24
62	83	33	54	4	22
61	82	32	53	3	21
60	81	31	52	2	14
59	80	30	51	1	7
58	79	29	49	0	0
57	78	28	48		

Examination June 2000
Math A

PART I

Answer all questions in this part. Each correct answer will receive 2 credits. No partial credit will be allowed. Record your answers in the spaces provided. [40]

1 Which inequality is represented in the graph below?

(1) $-4 < x < 2$ (3) $-4 < x \le 2$

(2) $-4 \le x < 2$ (4) $-4 \le x \le 2$ 1 _____

2 Which geometric figure has one and only one line of symmetry?

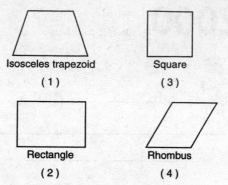

Isosceles trapezoid
(1)

Square
(3)

Rectangle
(2)

Rhombus
(4)

2 _____

3 Which number is rational?

(1) π (3) $\sqrt{7}$

(2) $\dfrac{5}{4}$ (4) $\sqrt{\dfrac{3}{2}}$

3 _____

4 Two numbers are in the ratio 2:5. If 6 is subtracted from their sum, the result is 50. What is the larger number?

(1) 55   (3) 40
(2) 45   (4) 35

4 _____

5 The quotient of $-\dfrac{15x^8}{5x^2}$, $x \neq 0$, is

(1) $-3x^4$ (3) $-3x^6$
(2) $-10x^4$ (4) $-10x^6$

5 _____

6 What is the inverse of the statement "If it is sunny, I will play baseball"?

(1) If I play baseball, then it is sunny.
(2) If it is not sunny, I will not play baseball.
(3) If I do not play baseball, then it is not sunny.
(4) I will play baseball if and only if it is sunny.   6 _____

7 Which ordered pair is the solution of the following system of equations?

$$3x + 2y = 4$$
$$-2x + 2y = 24$$

(1) $(2,-1)$ (3) $(-4,8)$
(2) $(2,-5)$ (4) $(-4,-8)$ 7 _____

8 Which equation represents a circle whose center is $(3,-2)$?

(1) $(x + 3)^2 + (y - 2)^2 = 4$
(2) $(x - 3)^2 + (y + 2)^2 = 4$
(3) $(x + 2)^2 + (y - 3)^2 = 4$
(4) $(x - 2)^2 + (y + 3)^2 = 4$ 8 _____

9 The set of integers {3,4,5} is a Pythagorean triple. Another such set is

(1) {6,7,8}   (3) {6,12,13}
(2) {6,8,12}   (4) {8,15,17}   9 _____

10 A truck travels 40 miles from point A to point B in exactly 1 hour. When the truck is halfway between point A and point B, a car starts from point A and travels at 50 miles per hour. How many miles has the car traveled when the truck reaches point B?

(1) 25   (3) 50
(2) 40   (4) 60   10 _____

11 If $a \neq 0$ and the sum of x and $\frac{1}{a}$ is 0, then

(1) $x = a$ (3) $x = -\frac{1}{a}$

(2) $x = -a$ (4) $x = 1 - a$ 11 _____

12 The accompanying figure shows the graph of the equation $x = 5$.

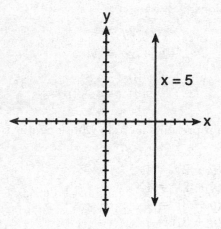

What is the slope in the line $x = 5$?
(1) 5   (3) 0
(2) −5   (4) undefined   12 _____

13 Which transformation does *not* always produce an image that is congruent to the original figure?

(1) translation   (3) rotation
(2) dilation   (4) reflection   13 _____

14 If rain is falling at the rate of 2 inches per hour, how many inches of rain will fall in x minutes?

(1) $2x$

(3) $\dfrac{60}{x}$

(2) $\dfrac{30}{x}$

(4) $\dfrac{x}{30}$

14 _____

15 The expression $(x - 6)^2$ is equivalent to
(1) $x^2 - 36$
(3) $x^2 - 12x + 36$
(2) $x^2 + 36$
(4) $x^2 + 12x + 36$

15 _____

16 How many different five-digit numbers can be formed from the digits 1, 2, 3, 4, and 5 if each digit is used only once?
(1) 120
(3) 24
(2) 60
(4) 20

16 _____

17 For five algebra examinations, Maria has an average of 88. What must she score on the sixth test to bring her average up to exactly 90?
(1) 92
(3) 98
(2) 94
(4) 100

17 _____

18 The graphs of the equations $y = x^2 + 4x - 1$ and $y + 3 = x$ are drawn on the same set of axes. At which point do the graphs intersect?
(1) $(1,4)$
(3) $(-2,1)$
(2) $(1,-2)$
(4) $(-2,-5)$

18 _____

19 If $2x^2 - 4x + 6$ is subtracted from $5x^2 + 8x - 2$, the difference is

(1) $3x^2 + 12x - 8$ (3) $3x^2 + 4x + 4$

(2) $-3x^2 - 12x + 8$ (4) $-3x^2 + 4x + 4$ 19 ____

20 What is the value of 3^{-2}?

(1) $\dfrac{1}{9}$ (3) 9

(2) $-\dfrac{1}{9}$ (4) -9 20 ____

PART II

Answer all questions in this part. Each correct answer will receive 2 credits. Clearly indicate the necessary steps, including appropriate formula substitutions, diagrams, graphs, charts, etc. For all questions in this part, a correct numerical answer with no work shown will receive only 1 credit. [10]

21 The formula for changing Celsius (C) temperature to Fahrenheit (F) temperature is $F = \dfrac{9}{5}C + 32$. Calculate, to the *nearest degree*, the Fahrenheit temperature when the Celsius temperature is -8.

22 Using only a ruler and compass, construct the bisector of angle *BAC* in the accompanying diagram.

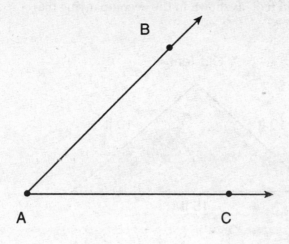

23 All seven-digit telephone numbers in a town begin with 245. How many telephone numbers may be assigned in the town if the last four digits do *not* begin or end in a zero?

24 The Rivera family bought a new tent for camping. Their old tent had equal sides of 10 feet and a floor width of 15 feet, as shown in the accompanying diagram.

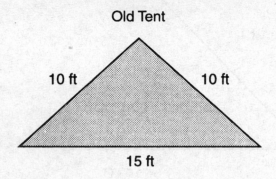

Old Tent

10 ft   10 ft

15 ft

If the new tent is similar in shape to the old tent and has equal sides of 16 feet, how wide is the floor of the new tent?

25 The accompanying graph represents the yearly cost of playing 0 to 5 games of golf at the Shadybrook Golf Course. What is the total cost of joining the club and playing 10 games during the year?

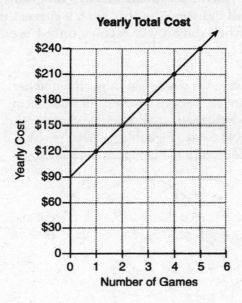

PART III

Answer all questions in this part. Each correct answer will receive 3 credits. Clearly indicate the necessary steps, including appropriate formula substitutions, diagrams, graphs, charts, etc. For all questions in this part, a correct numerical answer with no work shown will receive only 1 credit. [15]

26 The accompanying Venn diagram shows the number of students who take various courses. All students in circle *A* take mathematics. All in circle *B* take science. All in circle *C* take technology. What percentage of the students take mathematics or technology?

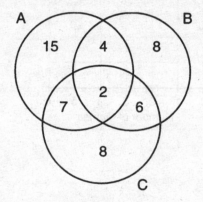

27 Hersch says if a triangle is an obtuse triangle, then it cannot also be an isosceles triangle. Using a diagram, show that Hersch is incorrect, and indicate the measures of all the angles and sides to justify your answer.

28 Tamika has a hard rubber ball whose circumference measures 13 inches. She wants to box it for a gift but can only find cube-shaped boxes of sides 3 inches, 4 inches, 5 inches, or 6 inches. What is the *smallest* box that the ball will fit into with the top on?

29 The distance from Earth to the imaginary planet Med is 1.7×10^7 miles. If a spaceship is capable of traveling 1,420 miles per hour, how many days will it take the spaceship to reach the planet Med? Round your answer to the *nearest day*.

30 A surveyor needs to determine the distance across the pond shown in the accompanying diagram. She determines that the distance from her position to point P on the south shore of the pond is 175 meters and the angle from her position to point X on the north shore is 32°. Determine the distance, PX, across the pond, rounded to the *nearest meter*.

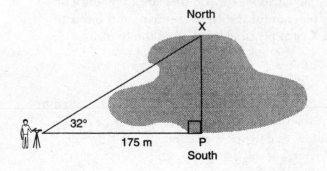

PART IV

Answer all questions in this part. Each correct answer will receive 4 credits. Clearly indicate the necessary steps, including appropriate formula substitutions, diagrams, graphs, charts, etc. For all questions in this part, a correct numerical answer with no work shown will receive only 1 credit. [20]

31 The owner of a movie theater was counting the money from 1 day's ticket sales. He knew that a total of 150 tickets were sold. Adult tickets cost $7.50 each and children's tickets cost $4.75 each. If the total receipts for the day were $891.25, how many of *each* kind of ticket were sold?

32 A treasure map shows a treasure hidden in a park near a tree and a statue. The map indicates that the tree and the statue are 10 feet apart. The treasure is buried 7 feet from the base of the tree and also 5 feet from the base of the statue. How many places are possible locations for the treasure to be buried? Draw a diagram of the treasure map, and indicate with an **X** *each* possible location of the treasure.

33 The scores on a mathematics test were 70, 55, 61, 80, 85, 72, 65, 40, 74, 68, and 84. Complete the accompanying table, and use the table to construct a frequency histogram for these scores.

Score	Tally	Frequency
40–49		
50–59		
60–69		
70–79		
80–89		

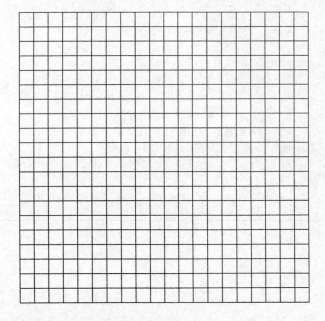

34 Paul orders a pizza. Chef Carl randomly chooses two different toppings to put on the pizza from the following: pepperoni, onion, sausage, mushrooms, and anchovies. If Paul will not eat pizza with mushrooms, determine the probability that Paul will *not* eat the pizza Chef Carl has made.

35 The area of the rectangular playground enclosure at South School is 500 square meters. The length of the playground is 5 meters longer than the width. Find the dimensions of the playground, in meters. [*Only an algebraic solution will be accepted.*]

Answers June 2000

Math A

Answer Key

PART I

1. 2	**6.** 2	**11.** 3	**16.** 1
2. 1	**7.** 3	**12.** 4	**17.** 4
3. 2	**8.** 2	**13.** 2	**18.** 4
4. 3	**9.** 4	**14.** 4	**19.** 1
5. 3	**10.** 1	**15.** 3	**20.** 1

PART II

21. 18

22. *See* Answers Explained section.

23. 8,100

24. 24′

25. $390

PART III

26. 84%

27. *See* Answers Explained section.

28. *See* Answers Explained section

29. 499 days

30. 109 meters $32° = \dfrac{y}{175}$.

PART IV

31. 65 adult tickets and 85 student tickets
32. *See* Answers Explained section.
33. *See* Answers Explained section.
34. *See* Answers Explained section.
35. Width = 20   Length = 25

Answers Explained

PART I

1. The accompanying graph includes −4 and all values between −4 to 2. The graph does not include 2, since there is an open circle around it.

The set of all values of x from −4 up to but not including 2 is represented by the inequality $-4 \leq x < 2$.

The correct choice is **(2)**.

2. A line of symmetry divides a figure into two mirror-image parts. If the figure is folded along the line of symmetry, the two parts will exactly coincide. To identify the geometric figure that has one and only one line of symmetry, consider each choice in turn:

- Choice (1): An isosceles trapezoid has exactly one line of symmetry, as shown in the accompanying figure.

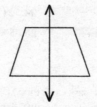

- Choice (2): A rectangle has two lines of symmetry, as shown in the accompanying figure.

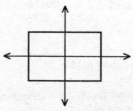

- Choice (3): A square has four lines of symmetry, as shown in the accompanying figure.

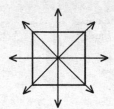

- Choice (4): A rhombus has two lines of symmetry, as shown in the accompanying figure.

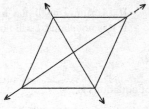

The correct choice is (1).

3. A number is rational if it can be written as the quotient of two integers, $\frac{x}{y}$, where $y \neq 0$. An expression that cannot be represented by a rational number is irrational. Hence, π, $\sqrt{7}$, and $\sqrt{\frac{3}{2}}$ are irrational numbers. Only $\frac{5}{4}$ is a rational number.

The correct choice is (2).

4. If two numbers are in the ratio 2:5, then the smaller of these two numbers can be represented by $2x$ and the larger by $5x$. If 6 is subtracted from their sum ($2x + 5x - 6$), the result is 50. Thus:

$$2x + 5x - 6 = 50$$
$$7x - 6 = 50$$
$$7x = 56$$

$$x = \frac{56}{7}$$

$$= 8$$

To find the value of the larger number, $5x$, substitute 8 for x:

$$5x = 5(8) = 40$$

The correct choice is (3).

5. The given quotient is:
$$-\frac{15x^8}{5x^2} \qquad (x \neq 0)$$

Divide the numerical coefficients:
Divide powers of the same base, x,
by subtracting exponents:

$$-\frac{15x^8}{5x^2} = \frac{\overset{3}{\cancel{15}}x^8}{\cancel{5}x^2}$$
$$= -3x^{8-2}$$
$$= -3x^6$$

The correct choice is **(3)**.

6. To form the inverse (negation) of a conditional statement of the form "If statement 1, then statement 2," negate each part by inserting the word *not*.

Given statement: If it is sunny, I will play baseball.
Inverse: If it is not sunny, I will not play baseball.

The correct choice is **(2)**.

7. The given system of equations is:

$$3x + 2y = 4$$
$$-2x + 2y = 24$$

Solve the system of equations algebraically or test each of the four answer choices until you find the ordered pair that makes each equation a true statement.

Method I: Solve the system of equations algebraically.
Eliminate y by multiplying both sides of
the second equation by -1:

$$-2x + 2y = 24$$
$$-1(-2x + 2y) = -1(24)$$
$$2x - 2y = -24$$

Add the result to the first equation:

$$3x + 2y = 4$$
$$+ 2x - 2y = -24$$
$$\overline{5x + 0 = -20}$$

Solve for x:

$$x = \frac{-20}{5}$$
$$= -4$$

Substitute -4 for x in the first (or second) equation:

$$3x + 2y = 24$$
$$3(-4) + 2y = 4$$
$$-12 + 2y = 4$$
$$\underline{+12 \qquad = +12}$$
$$2y = 16$$

Solve for y:
$$y = \frac{16}{2}$$
$$= 8$$

The solution to the given system of equations is the ordered pair $(-4, 8)$.
The correct choice is **(3)**.

<u>Method II: Work back from the answer choices.</u>
For each answer choice substitute the given values of x and y into each of the original equations. Continue until you find the ordered pair that satisfies both equations simultaneously. You should verify that the ordered pair $(-4, 8)$ works in both equations.
Let $x = -4$ and $y = 8$:

$$3x + 2y = 4 \qquad\qquad -2x + 2y = 24$$
$$3(-4) + 2(8) = 4 \qquad\qquad -2(-4) + 2(8) = 24$$
$$-12 + 16 = 4 \qquad\qquad 8 + 16 = 24$$

The correct choice is **(3)**.

8. An equation of a circle with center at point (h, k) and radius r is:

$$(x - h)^2 + (y - k)^2$$

Since the given circle has a center at point $(3, -2)$, $h = 3$ and $k = -2$. Also, since all four answer choices give $r^2 = 4$, the radius of the circle is 2. Thus, the equation of the given circle is:

$$(x - 3)^2 + (y - (-2))^2 = 2^2$$
$$(x - 3)^2 + (y + 2)^2 \quad = 4$$

The correct choice is **(2)**.

9. Since the given set of numbers, {3, 4, 5}, is a Pythagorean triple, the square of the largest number is equal to the sum of the squares of the other two: $3^2 + 4^2 = 5^2$. Test the three numbers in each of the answer choices until you find the choice that satisfies the Pythagorean relationship.

- Choice (1): $\overbrace{6^2 + 7^2}^{36\ +\ 49} \neq \overbrace{8^2}^{64}$. Since $36 + 49 = 64$, {6, 7, 8} is not a Pythagorean triple.

- Choice (2): $\overbrace{6^2 + 8^2}^{36\ +\ 64} \neq \overbrace{12^2}^{144}$. Since $36 + 64 = 144$, {6, 8, 12} is not a Pythagorean triple.

- Choice (3): $\overbrace{6^2 + 12^2}^{36\ +\ 144} \neq \overbrace{13^2}^{169}$. Since $36 + 144 = 169$, {6, 12, 13} is not a Pythagorean triple.

- Choice (4): $\overbrace{8^2 + 15^2}^{64\ +\ 225} = \overbrace{17^2}^{289}$. Since $64 + 225 = 289$, {8, 15, 17} is a Pythagorean triple.

The correct choice is **(4)**.

10. It is given that a truck travels 40 miles from point A to point B in exactly 1 hour. Halfway between point A and point B, the truck has traveled $\frac{1}{2}$ hour. At this time, a car starts from point A and travels 50 miles per hour. The truck reaches point B $\frac{1}{2}$ hour after the car starts from Point A. Thus, the distance the car travels in this amount of time is obtained by multiplying its rate of speed by the time traveled:

$$\text{Distance} = \text{Rate} \times \text{Time}$$

$$= 50\,\frac{\text{miles}}{\text{hour}} \times \frac{1}{2}\,\text{hour}$$

$$= \frac{\overset{25}{\cancel{50}}\ \text{miles}}{\cancel{\text{hour}}} \times \frac{1}{\cancel{2}}\,\cancel{\text{hour}}$$

$$= 25\,\text{miles}$$

The correct choice is **(1)**.

11. It is given that $a \neq 0$ and the sum of x and $\frac{1}{a}$ is 0; thus, $x + \frac{1}{a} = 0$. Subtracting $\frac{1}{a}$ from each side of the equation gives $x = -\frac{1}{a}$.

The correct choice is **(3)**.

12. The slope of any vertical line is undefined. Since the graph of the equation $x = 5$ is a vertical line, its slope is undefined.

The correct choice is **(4)**.

13. A translation, a rotation, and a reflection are transformations that do not affect the size and shape of a figure, so these transformations produce images that are always congruent to the original figure. Since a dilation may enlarge or shrink a figure, it does not always produce an image that is congruent to the original figure.

The correct choice is **(2)**.

14. It is given that rain is falling at the rate of 2 inches per hour. To find the number of inches of rain that will fall in x minutes, first convert the rate of rainfall to inches per minute:

$$2 \, \frac{\text{inches}}{\text{hour}} = \frac{2 \text{ inches}}{\text{hour}} \times \frac{1 \text{ hour}}{60 \text{ minutes}} = \frac{2}{60} \, \frac{\text{inches}}{\text{minutes}} = \frac{1}{30} \, \frac{\text{inches}}{\text{minutes}}$$

Multiply the rate of rain by x minutes:

$$\frac{1}{30} \, \frac{\text{inches}}{\text{minutes}} \cdot x \text{ minutes} = \frac{x}{30} \text{ inches}$$

The correct choice is **(4)**.

15. The given expressions is: $(x - 6)^2$

Rewrite the square of the binomial as a product of two binomials:

$$(x - 6)(x - 6)$$

Use FOIL to multiply the binomials together:

$$
\begin{aligned}
(x - 6)(x - 6) &= \overbrace{x \cdot x}^{\text{First}} + \overbrace{x(-6)}^{\text{Outer}} + \overbrace{(-6)x}^{\text{Inner}} + \overbrace{(-6)(-6)}^{\text{Last}} \\
&= x^2 - 6x - 6x + 36 \\
&= x^2 - 12x + 36
\end{aligned}
$$

The correct choice is **(3)**.

16. It is given that each of the digits 1, 2, 3, 4, and 5 is used only once to form a five-digit number. To find the total number of five-digit numbers that can be formed, use the counting principle:

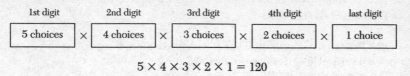

$$5 \times 4 \times 3 \times 2 \times 1 = 120$$

The correct choice is (**1**).

17. It is given that for five algebra examinations, Maria has an average score of 88, so her total score for the five tests is $88 \times 5 = 440$. Let x represent the score that Maria needs on the sixth test to bring her average up to exactly 90. Her average for the six tests is:

$$\frac{440 + x}{6} = 90$$

$$\cancel{6}\left(\frac{440 + x}{\cancel{6}}\right) = 6(90)$$

$$440 + x = 540$$
$$x = 540 - 440$$
$$= 100$$

The correct choice is (**4**).

18. It is given that the graphs of the equations $y = x^2 + 4x - 1$ and $y + 3 = x$ are drawn on the same set of axes. To determine the point at which the graphs intersect, consider each answer choice to find the ordered pair that satisfies both equations.

- Choice (1): $(1,4)$ does *not* satisfy $y + 3 = x$, since $4 + 3 \neq 1$
- Choice (2): $(1,-2)$ does *not* satisfy $y = x^2 + 4x - 1$, since $-2 = 1^2 + 4(1) - 1$
- Choice (3): $(-2,1)$ does *not* satisfy $y + 3 = x$, since $1 + 3 \neq -2$
- Choice (4): $(-2,-5)$ satisfies both equations:

$$\underline{y + 3 = x} \qquad\qquad \underline{y = x^2 + 4x - 1}$$

$$5 \mid 3 \overset{?}{=} -2 \qquad\qquad -5 \overset{?}{=} (-2)^2 + 4(-2) - 1$$

$$-2 \overset{?}{=} -2 \qquad\qquad -5 \overset{?}{=} 4 \qquad -8 \qquad -1$$

$$\qquad\qquad\qquad\qquad -5 \overset{?}{=} -5$$

The correct choice is **(4)**.

19. To subtract $2x^2 - 4x + 6$ from $5x^2 + 8x - 2$, write the polynomial to be subtracted underneath the other polynomial with like terms aligned in the same column:

$$5x^2 + 8x - 2$$
$$-\qquad\qquad$$
$$\underline{2x^2 - 4x + 6}$$

Change to an equivalent addition problem by reversing the signs of each term of the polynomial to be subtracted:

Add like terms in each column:

$$5x^2 + 8x - 2$$
$$+\qquad\qquad$$
$$\underline{-2x^2 + 4x - 6}$$
$$3x^2 + 12x - 8$$

The correct choice is **(1)**.

20. To evaluate the given expression, 3^{-2}, invert the base and make the exponent positive:

$$3^{-2} = \frac{1}{3^2} = \frac{1}{3 \times 3} = \frac{1}{9}$$

The correct choice is **(1)**.

PART II

21. The formula for changing Celsius (C) temperature to Fahrenheit (F) temperature is given as $F = \frac{9}{5}C + 32$. To calculate the Fahrenheit temperature when the Celsius temperature is -8, let $C = -8$, and solve for F:

Since $\frac{9}{5} = 1.8$ $\qquad\qquad F = \frac{9}{5}C \qquad +32$

and $\quad C = -8$ $\qquad\qquad\quad = 1.8(-8) \quad + 32$

$\qquad\qquad\qquad\qquad\qquad = -14.4 \quad + 32$

$\qquad\qquad\qquad\qquad\qquad = 17.6$

To the nearest degree, the equivalent Fahrenheit temperature is **18**.

22.

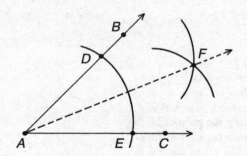

To construct the bisector of $\angle BAC$, shown in the accompanying diagram, follow these steps:

Step 1: Place your compass point at A and swing an arc intersecting $\overline{AB}$ at D and $\overline{AC}$ at E.

Step 2: Place your compass point at D and swing an arc.

Step 3: Using the same compass setting, place your compass point at E and swing an arc. Label the point at which the two arcs intersect as point F.

Step 4: Using a straightedge, draw $\overline{AF}$, the bisector of $\angle BAC$.

23. It is given that all seven-digit telephone numbers in a town begin with 245, and the last four digits do not begin or end in a zero:

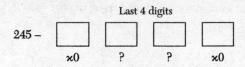

Last 4 digits

245 – [ ] [ ] [ ] [ ]
     x0   ?   ?   x0

To determine how many telephone numbers may be assigned in the town, find the number of digits that can be used to fill each position of the last four digits.

- The first or last positions may be filled by any one of the nine digits from 1 to 9:

245 – [9] [ ] [ ] [9]

- The remaining two positions may be filled by any one of the 10 digits from 0 to 9:

245 – [9] [10] [10] [9]

- Use the counting principle to determine the number of different ways in which the last four digits can be filled:

$$[9] \times [10] \times [10] \times [9] = 8{,}100$$

Hence, **8,100** telephone numbers may be assigned in the town.

24.

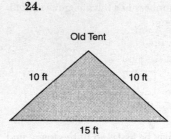

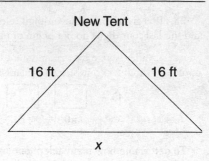

It is given that the Rivera family's old tent had equal sides of 10 feet and a floor of width 15 feet, and that the new tent is similar in shape with equal sides of 16 feet. The two tents are shown in the accompanying diagram. Since the lengths of corresponding sides of similar triangles are in proportion, let x represent the width of the floor of the new tent.

Set up a proportion:

$$\frac{\text{Old tent}}{\text{New tent}} = \frac{10}{16} = \frac{15}{x}$$

Cross-multiply:

$$10x = 15 \times 16$$

Solve for x:

$$x = \frac{240}{10} = 24$$

The floor of the new tent is **24 feet**.

25. The yearly cost of playing 0 to 5 games of golf is shown in the accompanying graph.

- The total cost of joining the golf club is the y-intercept of the graph, since at this point the number of games is 0. Hence, the cost of joining the club is $90.
- Since the yearly cost increases from $90 to $120 when the number of games increases from 0 to 1, the cost of playing one game is $120 − $90 = $30. Hence, the cost of playing 10 games is $30 × 10 = $300.
- The total cost of joining the club and playing 10 games during the year is, therefore, $90 + $300 = **$390**.

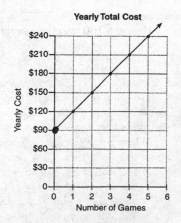

PART III

26. In the accompanying Venn diagram, circle A represents all students who take mathematics, circle B shows all students who take science, and circle C shows all students who take technology.

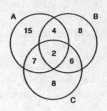

- The total number of students in the three courses is:

$$15 + 4 + 2 + 7 + 8 + 6 + 8 = 50$$

- Since the part of circle B that does not overlap circle A or circle C contains exactly 8 students, 8 students take only science. Hence, $50 - 8 = 42$ students take mathematics or technology.
- Because 42 of the 50 students take mathematics or technology, the percentage of the students in this category is:

$$\frac{42}{50} \times 100\% = 84\%$$

Thus, **84%** of the students take mathematics or technology.

27. According to Hersch, if a triangle is an obtuse triangle, then it cannot also be an isosceles triangle.

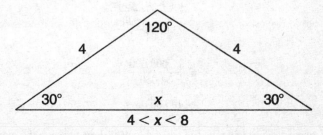

Hersch is incorrect, as shown by the accompanying diagram. The largest angle is 120°, and two of the sides are congruent, with a length of 4. Hence, the triangle is both obtuse and isosceles.

When indicating the measures of the angles and sides of this triangle, keep in mind that:

- The sum of the measures of the three angles must add to 180.
- The congruent sides must form the obtuse angle, so that the acute angles opposite these sides have equal measures.
- The side opposite the obtuse angle is the longest side of the triangle, and its length must be less than the sum of the lengths of the other two sides.

28. Tamika needs to know the diameter of the ball to determine what size box it will fit into. It is given that the hard rubber ball has a circumference of 13 inches.

Since circumference $= \pi \times$ diameter, the diameter is $\dfrac{13}{\pi} \approx \dfrac{13}{3.14} \approx 4.14$ inches.

Because the diameter is between 4 and 5 inches, the smallest cube-shaped box that the ball will fit into with the top on is one with a side length of 5 inches.
Tamika must use a **5-inch box**.

29. It is given that the distance from Earth to the imaginary planet Med is 1.7 $\times 10^7$ miles, and that a spaceship is traveling there at 1,420 miles per hour. To find the number of days it will take to reach the planet, first convert the speed to miles per day:

$$1,420 \; \frac{\text{miles}}{\text{hour}} = 1,420 \; \frac{\text{miles}}{\cancel{\text{hour}}} \times \frac{24 \; \cancel{\text{hours}}}{\text{day}}$$

$$= 34,080 \; \frac{\text{miles}}{\text{day}}$$

Since rate times time equals distance:

$$\text{Time} = \frac{\text{Distance}}{\text{Rate}}$$

$$= \frac{1.7 \times 10^7 \; \cancel{\text{miles}}}{34,080 \; \dfrac{\cancel{\text{miles}}}{\text{day}}}$$

$$= \frac{17,000,000}{34,080} \; \text{days}$$

$$\approx 498.83 \; \text{days}$$

To the nearest day, it will take the spaceship **499 days** to reach the planet Med.

30. It is given that a surveyor needs to determine the distance, PX, across a pond as shown in the accompanying diagram. Use the tangent ratio to find PX:

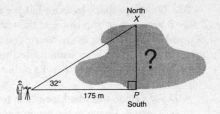

$$\tan 32° = \frac{PX}{175}$$

$$0.6249 = \frac{PX}{175}$$

$$PX = 0.6249 \times 175$$
$$= 109.36$$

To the nearest meter, the distance across the pond is **109 meters**.

PART IV

31. It is given that 150 movie tickets were sold, adult tickets cost $7.50 each, children's tickets cost $4.75 each, and the total receipts for the day were $891.25. If x represents the number of adult tickets sold, $150 - x$ represents the number of children's tickets sold. Hence, total receipts are:

$$\$7.50x + \$4.75(150 - x) = \$891.25$$
Simplify:
$$\$7.50x + \$712.50 - \$4.75x = \$891.25$$
$$\$2.75x + \$712.50 = \$891.25$$
$$-\$712.50 = -\$712.50$$
$$\overline{\$2.75x \qquad\qquad = \$178.75}$$

Solve for x to find the number of adult tickets:
$$x = \frac{\$178.75}{\$2.75}$$
$$= 65$$

Substitute for x to find the number of children's tickets: $150 - x = 150 - 65 = 85$

Thus, **65 adult tickets** and **85 children's tickets** were sold.

32. It is given that a treasure hidden in a park is located 7 feet from a tree and 5 feet from a statue. The tree and the statue are 10 feet apart. To find the number of possible locations for the treasure, consider this a locus problem. The locus of all points at a given distance from a fixed point is a circle with the fixed point as its center and the given distance as its radius:

- Draw a circle with the base of the tree as its center and with the radius length of 7 feet, as shown in the accompanying diagram.

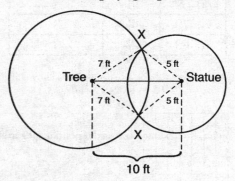

- Draw a circle with the base of the statue as its center and with radius length of 5 feet, as shown in the accompanying figure.
- Because $7 + 5 > 10$, the circles will intersect in two points. Mark each point of intersection with an **X** to indicate the possible locations for the treasure.

Thus, **2 points** are possible locations for the hidden treasure.

33. It is given that the scores on a mathematics test were 70, 55, 61, 80, 85, 72, 65, 40, 74, 68, and 84. Organize the data in the table below:

Score	Tally	Frequency
40–49	I	1
50–59	I	1
60–69	III	3
70–79	III	3
80–89	III	3

Use the data in the table to draw a histogram, as shown in the accompanying figure.

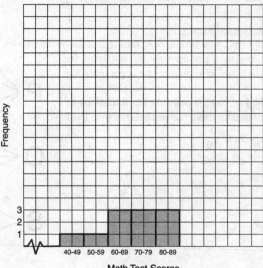

Math Test Scores

34. It is given that Paul orders a pizza and that Chef Carl randomly chooses two different toppings from the following five toppings: pepperoni, onion, sausage, mushrooms, and anchovies. Paul will *not* eat pizza with mushrooms.

Find the probability that Paul will *not* eat the pizza by determining the probability that one of the toppings chosen by Carl will be mushrooms.

<u>Method I: Use combinations.</u>
- Since order doesn't matter, two toppings can be randomly selected from five toppings in $_5C_2$ different ways.
- A mushroom topping and any of the four remaining toppings can be randomly selected in $_1C_1 \cdot {}_4C_1$ different ways.

- P(Paul will *not* eat pizza) $= \dfrac{_1C_1 \cdot {}_4C_1}{_5C_2} = \dfrac{1 \times 4}{10} = \dfrac{4}{10}$

The probability that Paul will *not* eat the pizza is $\dfrac{4}{10}$.

<u>Method II: Draw a diagram.</u>
Let P = pepperoni, O= onion, S = sausage, and A = anchovies. Then the possible combinations of toppings can be obtained by drawing a tree diagram, as shown in the accompanying figure.

Since exactly 4 of the 10 possible combinations of toppings contain mushrooms, the probability that Paul will *not* eat the pizza is $\dfrac{4}{10}$.

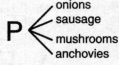

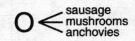

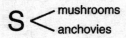

35. It is given that the area of the rectangular playground enclosure at South School is 500 square meters, and that the length is 5 meters longer than the width. The dimensions are shown in the accompanying diagram.

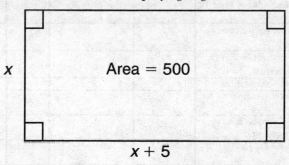

Let x represent the width of the playground in meters. Then $x + 5$ is the length of the playground in meters. Since the area of a rectangle is length times width:

$$x(x + 5) = 500$$

Solve for x:
$$x^2 + 5x = 500$$
$$x^2 + 5x - 500 = 0$$
$$(x - 20)(x + 25) = 0$$
$$x - 20 = 0 \quad or \quad x + 25 = 0$$
$$x = 20 \quad or \quad x = -25$$

↑ Reject since width cannot be a negative number.

Hence, the width of the playgroud, x, is **20 meters** and the length, $x + 5$, is **25 meters**.

Topic	Question Numbers	Number of Points	Your Points	Your Percentage
1. Numbers; Properties of Real Numbers; Percent	3	2		
2. Operations on Rat'l. Numbers & Monomials	5	2		
3. Laws of Exponents for Integer Exponents; Scientific Notation	20	2		
4. Operations on Polynomials	19	2		
5. Square Root; Operations with Radicals	—	—		
6. Evaluating Formulas & Algebraic Expressions	21	2		
7. Solving Linear Eqs. & Inequalities	1	2		
8. Solving Literal Eqs. & Formulas for a Particular Letter	11	2		
9. Alg. Operations (including factoring)	15	2		
10. Solving Quadratic Eqs.; Graphs of Parabolas & Circles	8, 35	$2 + 4 = 6$		
11. Coordinate Geometry (graphs of linear eqs.; slope; midpoint; distance)	12	2		
12. Systems of Linear Eqs. & Inequalities (alg. & graph. solutions)	7	2		
13. Mathematical Modeling Using: Eqs.; Tables; Graphs of Linear Eqs.	25	2		
14. Linear-Quad Systems (alg. & graph. solutions)	18	2		
15. Word Problems Requiring Arith. or Alg. Reasoning	10, 29, 31	$2 + 3 + 4 = 9$		
16. Areas, Perims., Circums., Vols. of Common Figures	28	3		
17. Angle & Line Relationships (suppl., compl., vertical angles; parallel lines; congruence)	—	—		
18. Ratio & Proportion (incl. similar polygons)	4, 24	$2 + 2 = 4$		
19. Pythagorean Theorem	9	2		
20. Right Triangle Trig. & Indirect Measurement	30	3		

Topic	Question Numbers	Number of Points	Your Points	Your Percentage
21. Logic (symbolic rep.; conditionals; logically equiv. statements; valid arguments)	6	2		
22. Probability (incl. tree diagrams & sample spaces)	—	—		
23. Counting Methods & Sets (Venn diagrams)	26	3		
24. Permutations & Combinations	16, 23, 34	2 + 2 + 4 = 8		
25. Statistics (mean, percentiles, quartiles; freq. dist., histograms, stem & leaf plots; box-and-whiskers plots; circle graphs	17, 33	4		
26. Properties of Triangles & Parallelograms	27	3		
27. Transformations (reflections translations, rotations, dilations)	13	2		
28. Symmetry	2	2		
29. Area and Transformations using Coordinates	—	—		
30. Locus & Construction	22, 32	2 + 4 = 6		
31. Dimensional Analysis & Units of Measurement	14	2		

MAP TO LEARNING STANDARDS

Key Ideas	Item Numbers
Mathematical Reasoning	6, 26
Number and Numeration	3, 11, 28
Operations	2, 5, 13, 15, 19, 20, 21, 29, 31
Modeling/Multiple Representation	1, 14, 22, 27, 32
Measurement	4, 9, 10, 17, 24, 30, 33
Uncertainty	16, 23, 34
Patterns/Functions	7, 8, 12, 18, 25, 31, 35

HOW TO CONVERT YOUR RAW SCORE TO YOUR MATH A REGENTS EXAMINATION SCORE

Below is the conversion chart that must be used to determine your final score on the June 2000 Regents Examination in Math A. To find your final exam score, locate in the column labeled "Raw Score" the total number of points you scored out of a possible 85 points. Since partial credit is allowed in Parts II, III, and IV of the test, you may need to approximate the credit you would receive for a solution that is not completely correct. Then locate in the adjacent column to the right the scaled score that corresponds to your raw score. The scaled score is your final Math A Regents Examination score.

Regents Examination in Math A—June 2000
Chart for Converting Total Test Raw Scores to
Final Examination Scores (Scaled Scores)

Raw Score	Scaled Score	Raw Score	Scaled Score	Raw Score	Scaled Score
85	100	56	79	27	50
84	99	55	78	26	49
83	99	54	77	25	48
82	98	53	76	24	46
81	98	52	75	23	45
80	97	51	75	22	44
79	97	50	74	21	43
78	96	49	73	20	42
77	96	48	72	19	41
76	95	47	71	18	39
75	95	46	70	17	38
74	94	45	69	16	37
73	93	44	68	15	36
72	92	43	67	14	35
71	92	42	66	13	34
70	91	41	65	12	33
69	90	40	64	11	32
68	89	39	63	10	31
67	88	38	62	9	30
66	87	37	61	8	29
65	87	36	60	7	28
64	86	35	59	6	27
63	85	34	58	5	26
62	84	33	57	4	22
61	83	32	56	3	17
60	82	31	55	2	12
59	82	30	53	1	0
58	81	29	52	0	0
57	80	28	51		

Examination August 2000

Math A

PART I

Answer all questions in this part. Each correct answer will receive 2 credits. No partial credit will be allowed. Record your answers in the spaces provided. [40]

1 The product of $2x^3$ and $6x^5$ is

(1) $10x^8$ (3) $10x^{15}$

(2) $12x^8$ (4) $12x^{15}$ 1____

2 A hockey team played n games, losing four of them and winning the rest. The ratio of games won to games lost is

(1) $\dfrac{n-4}{4}$ (3) $\dfrac{4}{n}$

(2) $\dfrac{4}{n-4}$ (4) $\dfrac{n}{4}$ 2____

3 In the coordinate plane, what is the total number of points 5 units from the origin and equidistant from both the x- and y-axes?

(1) 1   (3) 0

(2) 2   (4) 4   3____

4 Expressed in decimal notation, 4.726×10^{-3} is

(1) 0.004726   (3) 472.6
(2) 0.04726   (4) 4,726   4 _____

5 Which table does *not* show an example of direct variation?

(1)
x	y
1	4
2	8
3	12
4	16

(3)
x	y
1	$\frac{1}{2}$
2	1
3	$\frac{3}{2}$
4	2

(2)
x	y
2	24
4	12
6	8
8	6

(4)
x	y
−4	−20
−3	−15
−2	−10
−1	−5

5 _____

6 If $a < b$, $c < d$, and a, b, c, and d are all greater than 0, which expression is always true?

(1) $a - c + b - d = 0$ (3) $\dfrac{a}{b} > \dfrac{b}{c}$
(2) $a + c > b + d$ (4) $ac < bd$ 6 _____

7 The volume of a cube is 64 cubic inches. Its total surface area, in square inches, is

(1) 16   (3) 96
(2) 48   (4) 576   7 _____

8 On an English examination, two students received scores of 90, five students received 85, seven students received 75, and one student received 55. The average score on this examination was

(1) 75   (3) 77
(2) 76   (4) 79   8 _____

9 Which equation represents a line parallel to the line $y = 2x - 5$?

(1) $y = 2x + 5$ (3) $y = 5x - 2$
(2) $y = -\frac{1}{2}x - 5$ (4) $y = -2x - 5$ 9 _____

10 The operation * for the set $\{p, r, s, v\}$ is defined in the accompanying table. What is the inverse element of r under the operation *?

*	p	r	s	v
p	s	v	p	r
r	v	p	r	s
s	p	r	s	v
v	r	s	v	p

(1) p (3) s
(2) r (4) v 10 _____

11 A box contains six black balls and four white balls. What is the probability of selecting a black ball at random from the box?

(1) $\frac{1}{10}$ (3) $\frac{4}{6}$
(2) $\frac{6}{10}$ (4) $\frac{6}{4}$ 11 _____

12 The solution set for the equation $x^2 - 2x - 15 = 0$ is

(1) $\{5,3\}$ (3) $\{-5,3\}$
(2) $\{5,-3\}$ (4) $\{-5,-3\}$ 12 _____

13 What is the value of y in the following system of equations?

$$2x + 3y = 6$$
$$2x + y = -2$$

(1) 1   (3) −3

(2) 2   (4) 4   13 _____

14 What is the converse of the statement "If it is sunny, I will go swimming"?

(1) If it is not sunny, I will not go swimming.

(2) If I do not go swimming, then it is not sunny.

(3) If I go swimming, it is sunny.

(4) I will go swimming if and only if it is sunny.   14 _____

15 Solve for x: $15x - 3(3x + 4) = 6$

(1) 1   (3) 3

(2) $-\frac{1}{2}$ (4) $\frac{1}{3}$ 15 _____

16 The expression $2\sqrt{50} - \sqrt{2}$ is equivalent to

(1) $2\sqrt{48}$ (3) $9\sqrt{2}$

(2) 10   (4) $49\sqrt{2}$ 16 _____

17 Which is an equation of the parabola shown in the accompanying diagram?

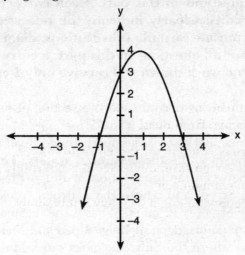

(1) $y = -x^2 + 2x + 3$ (3) $y = x^2 + 2x + 3$
(2) $y = -x^2 - 2x + 3$ (4) $y = x^2 - 2x + 3$ 17____

18 If two sides of a triangle are 1 and 3, the third side may be

(1) 5   (3) 3
(2) 2   (4) 4   18____

19 A girl can ski down a hill five times as fast as she can climb up the same hill. If she can climb up the hill and ski down in a total of 9 minutes, how many minutes does it take her to climb up the hill?

(1) 1.8   (3) 7.2
(2) 4.5   (4) 7.5   19____

20 When $3x^2 - 2x + 1$ is subtracted from $2x^2 + 7x + 5$, the result will be

(1) $-x^2 + 9x + 4$ (3) $-x^2 + 5x + 6$
(2) $x^2 - 9x - 4$ (4) $x^2 + 5x + 6$ 20____

PART II

Answer all questions in this part. Each correct answer will receive 2 credits. Clearly indicate the necessary steps, including appropriate formula substitutions, diagrams, graphs, charts, etc. For all questions in this part, a correct numerical answer with no work shown will receive only 1 credit.   [10]

21 The accompanying diagram shows a section of the city of Tacoma. High Road, State Street, and Main Street are parallel and 5 miles apart. Ridge Road is perpendicular to the three parallel streets. The distance between the intersection of Ridge Road and State Street and where the railroad tracks cross State Street is 12 miles. What is the distance between the intersection of Ridge Road and Main Street and where the railroad tracks cross Main Street?

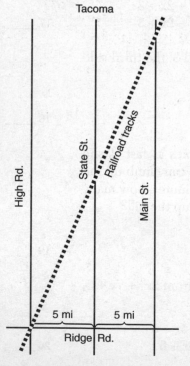

22 Perform the indicated operation and express the result in simplest terms:

$$\frac{x}{x+3} \div \frac{3x}{x^2-9}$$

23 Kerry is planning a rectangular garden that has dimensions of 4 feet by 6 feet. Kerry wants one-half of the garden to have roses, and she says that the rose plot will have dimensions of 2 feet by 3 feet. Is she correct? Explain.

24 The sum of the ages of the three Romano brothers is 63. If their ages can be represented as consecutive integers, what is the age of the middle brother?

25 Alan, Becky, Jesus, and Mariah are four students in the chess club. If two of these students will be selected to represent the school at a national convention, how many combinations of two students are possible?

PART III

Answer all questions in this part. Each correct answer will receive 3 credits. Clearly indicate the necessary steps, including appropriate formula substitutions, diagrams, graphs, charts, etc. For all questions in this part, a correct numerical answer with no work shown will receive only 1 credit.   [15]

26 John, Dan, Karen, and Beth went to a costume ball. They chose to go as Anthony and Cleopatra, and Romeo and Juliet. John got the costumes for Romeo and Cleopatra, but not his own costume. Dan saw the costumes for Juliet and himself. Karen went as Anthony. Beth drove two of her friends, who were dressed as Anthony and Cleopatra, to the ball. What costume did John wear?

27 To measure the length of a hiking trail, a worker uses a device with a 2-foot-diameter wheel that counts the number of revolutions the wheel makes. If the device reads 1,100.5 revolutions at the end of the trail, how many miles long is the trail, to the *nearest tenth of a mile?*

28 The coordinates of the endpoints of $\overline{AB}$ are $A(2,6)$ and $B(4,2)$. Is the image $\overline{A''B''}$ the same if it is reflected in the x-axis, then dilated by $\frac{1}{2}$ as the image is if it is dilated by $\frac{1}{2}$, then reflected in the x-axis? Justify your answer. [*The use of the accompanying grid is optional.*]

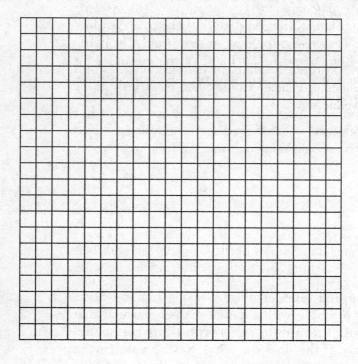

29 After an ice storm, the following headlines were reported in the *Glacier County Times*:

   Monday: Ice Storm Devastates County—8 out of every 10 homes lose electrical power

   Tuesday: Restoration Begins—Power restored to $\frac{1}{2}$ of affected homes

   Wednesday: More Freezing Rain—Power lost by 20% of homes that had power on Tuesday

   Based on these headlines, what fractional portion of homes in Glacier County had electrical power on Wednesday?

30 Katrina hikes 5 miles north, 7 miles east, and then 3 miles north again. To the *nearest tenth of a mile*, how far, in a straight line, is Katrina from her starting point?

PART IV

Answer all questions in this part. Each correct answer will receive 4 credits. Clearly indicate the necessary steps, including appropriate formula substitutions, diagrams, graphs, charts, etc. For all questions in this part, a correct numerical answer with no work shown will receive only 1 credit.   [20]

31 Mr. Santana wants to carpet exactly half of his rectangular living room. He knows that the perimeter of the room is 96 feet and that the length of the room is 6 feet longer than the width. How many square feet of carpeting does Mr. Santana need?

32 Ashanti is surveying for a new parking lot shaped like a parallelogram. She knows that three of the vertices of parallelogram $ABCD$ are $A(0,0)$, $B(5,2)$, and $C(6,5)$. Find the coordinates of point D and sketch parallelogram $ABCD$ on the accompanying set of axes. Justify mathematically that the figure you have drawn is a parallelogram.

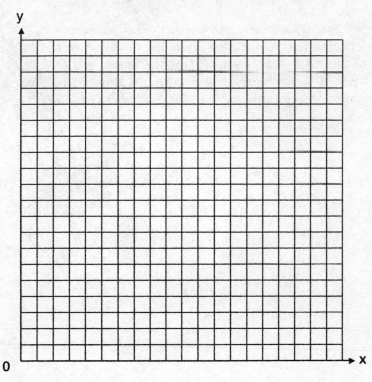

33 A 10-foot ladder is to be placed against the side of a building. The base of the ladder must be placed at an angle of 72° with the level ground for a secure footing. Find, to the *nearest inch*, how far the base of the ladder should be from the side of the building *and* how far up the side of the building the ladder will reach.

34 The telephone company has run out of seven-digit
telephone numbers for an area code. To fix this
problem, the telephone company will introduce a
new area code. Find the number of new seven-
digit telephone numbers that will be generated for
the new area code if both of the following conditions must be met:

- The first digit cannot be a zero or a one.
- The first three digits cannot be the emergency
number (911) or the number used for information (411).

35 Jack is building a rectangular dog pen that he wishes to enclose. The width of the pen is 2 yards less than the length. If the area of the dog pen is 15 square yards, how many yards of fencing would he need to completely enclose the pen?

Answers
August 2000
Math A

Answer Key

PART I

1. (2)	**6.** (4)	**11.** (2)	**16.** (3)
2. (1)	**7.** (3)	**12.** (2)	**17.** (1)
3. (4)	**8.** (4)	**13.** (4)	**18.** (3)
4. (1)	**9.** (1)	**14.** (3)	**19.** (4)
5. (2)	**10.** (4)	**15.** (3)	**20.** (1)

PART II

21. 24 miles

22. $\dfrac{x-3}{3}$

23. Kerry is incorrect. See *Answers Explained* section.

24. 21

25. 6

PART III

26. Juliet. See *Answers Explained* section.

27. 1.3

28. Yes. See *Answers Explained* section.

29. $\dfrac{48}{100}$

30. 10.6

PART IV

31. 283.5 or 284

32. $D(1,3)$ and accompanying sketch. See *Answers Explained* section.

33. 114 inches or 9 feet 6 inches, and 37 inches or 3 feet 1 inch

34. 7.98×10^6 or 7,980,000

35. 16

Answers Explained

PART I

1. To find the product of $2x^3$ and $6x^5$, multiply the numerical coefficients and then multiply the powers of x by adding the exponents. Thus:

$$2x^3 \cdot 6x^5 = (2 \cdot 6)\left(x^3 \cdot x^5\right)$$
$$= 12x^{3+5}$$
$$= 12x^8$$

The correct choice is (2).

2. It is given that of the n games played by a hockey team, four were lost and the rest won. Hence, the ratio of games won $(n - 4)$ to games lost (4) is the fraction

$$\frac{\text{games won}}{\text{games lost}} = \frac{n-4}{4}$$

The correct choice is (1).

3. The given problem is to find the total number of points that satisfy two conditions: (1) the points are 5 units from the origin and (2) the points are equidistant from both the x- and y-axes. First sketch the graphs that satisfy both conditions on the same set of axes. Then count the number of points at which the graphs intersect.

- Condition 1: The set of points 5 units from the origin can be represented by a circle whose center is at the origin and whose radius is 5 units, as shown in the accompanying diagram.

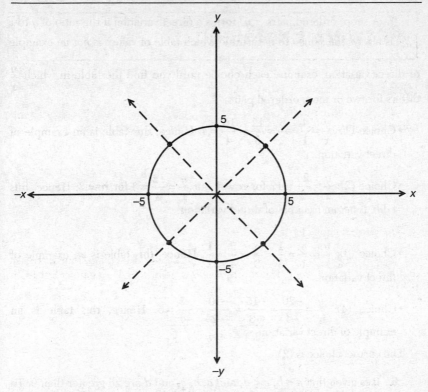

• Condition 2: The set of points equidistant from both the x- and y-axes can be represented by a pair of lines that bisect the angles formed by the coordinate axes, as shown in the same diagram.

• Since the graphs that represent the two conditions intersect at four different points, the total number of points that satisfy both conditions is 4.

The correct choice is (4).

4. Rewrite the given expression, 4.726×10^{-3}, so that it matches the form of the answer choices.

Change from a negative exponent to a positive exponent by inverting the base:

$$4.726 \times 10^{-3} = 4.726 \times \frac{1}{10^3}$$

$$= \frac{4.726}{1,000}$$

Divide by 1,000 by moving the decimal point of the numerator three places to the left:

$$= 0.004726$$

The correct choice is (1).

5. A set of ordered pairs (x,y) forms a direct variation if the ratio of y to x $\left(\dfrac{y}{x}\right)$ is always the same. To determine which table of values is *not* an example of direct variation, examine each choice until you find the table in which $\dfrac{y}{x}$ differs for two or more ordered pairs.

- Choice (1): $\dfrac{y}{x} = \dfrac{4}{1} = \dfrac{8}{2} = \dfrac{12}{3} = \dfrac{16}{4} = 4$. Hence, this table is an example of direct variation.

- Choice (2): $\dfrac{y}{x} = \dfrac{24}{2} = 12$ for row 1 but $\dfrac{y}{x} = \dfrac{12}{4} = 3$ for row 2. Hence, this table is *not* an example of direct variation.

- Choice (3): $\dfrac{y}{x} = \dfrac{\frac{1}{2}}{1} = \dfrac{1}{2} = \dfrac{\frac{3}{2}}{3} = \dfrac{2}{4} = \dfrac{1}{2}$. Hence, this table is an example of direct variation.

- Choice (4): $\dfrac{y}{x} = \dfrac{-20}{-4} = \dfrac{-15}{-3} = \dfrac{-10}{-2} = \dfrac{-5}{-1} = 5$. Hence, this table is an example of direct variation.

The correct choice is **(2)**.

6. It is given that $a < b$, $c < d$, and $a, b, c,$ and d are all greater than 0. To determine which expression is always true, pick easy numbers for $a, b, c,$ and d that satisfy the given conditions. For example, let $a = 1, b = 2, c = 3,$ and $d = 4$. Then plug these values in each of the given expressions until you find the choice that works.

- Choice (1): $a - c + b - d = 1 - 3 + 2 - 4 = -4$. Since $-4 < 0$, it is false that $a - c + b - d = 0$.

- Choice (2): $a + c = 1 + 3 = 4$ and $b + d = 2 + 4 = 6$. Since $4 < 6$, it is false that $a + c > b + d$.

- Choice (3): $\dfrac{a}{d} = \dfrac{1}{4}$ and $\dfrac{b}{c} = \dfrac{2}{3}$. Since $\dfrac{1}{4} < \dfrac{2}{3}$, it is false that $\dfrac{a}{d} > \dfrac{b}{c}$.

- Choice (4): $ac = 1 \cdot 3 = 3$ and $bd = 2 \cdot 4 = 8$. Since $3 < 8$, it is true that $ac < bd$.

The correct choice is **(4)**.

7. If the edge length of a cube is represented by s, then the volume of the cube is s^3 and the area of each side is s^2. Since a cube has 6 congruent sides, the total surface area of the cube is $6s^2$.

- It is given that the volume of a cube is 64 cubic inches. Thus, the edge length is 4 inches, since $4 \times 4 \times 4 = 64$.

- Since the edge length of the cube is 4 inches, the area of each of the six sides is $4 \times 4 = 16$ square inches.

- The total surface area of the cube is $6 \times 16 = 96$ square inches.

The correct choice is **(3)**.

8. The average of a set of scores is the sum of the scores divided by the total number of scores in the set. It is given that on an English examination, two students received scores of 90, five students received 85, seven students received 75, and one student received 55. Thus:

$$\text{average score} = \frac{\text{total number of points scored}}{\text{total number of test scores}}$$

$$= \frac{(2 \times 90) + (5 \times 85) + (7 \times 75) + (1 \times 55)}{2 + 5 + 7 + 1}$$

$$= \frac{180 + 425 + 525 + 55}{15}$$

$$= \frac{1{,}185}{15}$$

$$= 79$$

The correct choice is **(4)**.

9. If the equation of a line has the form $y = mx + b$, then m, the coefficient of x, represents the slope of the line and b represents the y-intercept. In the given equation, $y = 2x - 5$, $m = 2$, and $b = -5$. Thus, the slope of the line is 2.

Two lines are parallel if their slopes are equal. To determine which equation represents a line parallel to the given line, $y = 2x - 5$, evaluate each choice until you find an equation in which $m = 2$. Only choice (1), $y = 2x + 5$, has 2 as the coefficient of x. Since $m = 2$, for both lines, the two lines are parallel.

The correct choice is **(1)**.

10. The operation $*$ for the set $\{p,r,s,v\}$ is defined in the given table. To find the inverse element of r under the operation $*$, first find the identity element for $*$.

The identity element for an operation always returns the same element of the set as the one on which it operates. In the given table, notice that the row labeled s is the same as the row to the right of the operation symbol. Similarly, the vertical column labeled s is the same as the column under the operation symbol. This means that $s*x = x*s = x$ for $x = p, r, s,$ and v. Thus, s is the identity element.

*	p	r	s	v
p	s	v	p	r
r	v	p	r	s
s	p	r	s	v
v	r	s	v	p

The inverse of a particular element of a set is the member of the set that returns the identity as the result of the stated operation.

To find the inverse element of r, look across the row labeled r for the identity s, as shown in the accompanying table. Since $r*v = s$, v is the inverse element of r under the operation *.

*	p	r	s	v
p	s	v	p	r
$\longrightarrow r$	v	p	r	$\circledS$
s	p	r	s	v
v	r	s	v	p

The correct choice is **(4)**.

11. The probability of an event occurring can be obtained by dividing the number of favorable outcomes by the total number of possible outcomes. It is given that a box contains six black balls and four white balls. To find the probability of selecting a black ball at random from the box, form a fraction in which the numerator is the number of black balls (favorable outcomes) and the denominator is the total number of balls (possible outcomes). Thus:

$$P(\text{black ball}) = \frac{\text{number of black balls}}{\text{total number of balls}}$$
$$= \frac{6}{6+4}$$
$$= \frac{6}{10}$$

The correct choice is **(2)**.

12. The given expression, $x^2 - 2x - 15 = 0$, is a quadratic equation in which 0 appears alone on one side. Hence, the solution set can be determined by factoring the quadratic trinomial and then setting each factor equal to 0.

The given equation is: $x^2 - 2x - 15 = 0$

Factor the left side of the equation as
the product of two binomials: $(x + ?)(x + ?) = 0$

The missing terms of the binomial factors are the two integers whose sum is -2,
the coefficient of the x-term in $x^2 - 2x$
$- 15$, and whose product is -15, the constant term in $x^2 - 2x - 15$. Since $(-5) +$
$(+3) = -2$ and $(-5)(+3) = -15$, the miss-
ing terms are -5 and $+3$: $(x - 5)(x + 3) = 0$

If the product of two terms is 0, then
at least one of these terms is 0: $x - 5 = 0$ or $x + 3 = 0$

Solve each first-degree equation: $x = 5$ or $x = -3$

Write the solution set: $\{5, -3\}$

The correct choice is (2).

13. The given system of equations is:

$$2x + 3y = 6$$
$$2x + y = -2$$

Solve the system of equations algebraically by eliminating one of the variables. Since you are asked to find the value of y, eliminate x by subtracting the
second equation from the first:

$$2x + 3y = 6 \qquad\qquad 2x + 3y = 6$$

$-$ $\qquad\qquad \Rightarrow \qquad +$

$$\underline{2x + y\ = -2} \qquad\qquad \underline{-2x - y = +2}$$

$$2y = 8$$

$$y = \frac{8}{2} = 4$$

The correct choice is (4).

14. The converse of a conditional statement that has the form "If statement 1, then statement 2" is the conditional statement "If statement 2, then
statement 1."

The given statement is: If it is sunny, I will go swimming.

(statement 1: it is sunny; statement 2: I will go swimming.)

Form the converse by interchanging statements 1 and 2:

If I go swimming, it is sunny.

The correct choice is **(3)**.

15. The given equation is:

$$15x - 3(3x + 4) = 6$$

Remove the parentheses by
multiplying each term inside the
parentheses by the number in front of
the parentheses:

$$15x - 9x - 12 = 6$$

Combine like terms:

$$6x - 12 = 6$$

Isolate the variable term by adding
12 to each side of the equation:

$$\underline{+12 = +12}$$

$$6x = 18$$

Divide each side of the equation
by 6:

$$\frac{6x}{6} = \frac{18}{6}$$

$$x = 3$$

The correct choice is **(3)**.

16. To combine two unlike radicals, rewrite the radicals so that both terms
have the same radicand.

The given expression is:

$$2\sqrt{50} - \sqrt{2}$$

Rewrite 50 as the product of two
positive integers, one of which is the
greatest perfect square factor of 50:

$$2\sqrt{25 \cdot 2} - \sqrt{2}$$

Write the radical over each factor:

$$2 \cdot \sqrt{25} \cdot \sqrt{2} - \sqrt{2}$$

Let $\sqrt{25} = 5$:

$$2 \cdot 5\sqrt{2} - \sqrt{2}$$

Subtract like radicals by subtracting the numbers multiplying the
radicals:

$$10\sqrt{2} - 1\sqrt{2} = 9\sqrt{2}$$

The correct choice is **(3)**.

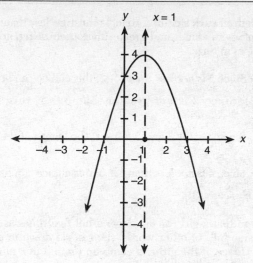

17. As shown in the accompanying diagram, the given parabola opens downward and has $x = 1$ as a vertical line of symmetry.

In general, the graph of an equation of the form $y = ax^2 + bx + c$ is a parabola. If $a < 0$, the parabola opens downward, and if $a > 0$ the parabola opens upward. An equation of the axis of symmetry of the parabola $y = ax^2 + bx + c$ is $x = -\dfrac{b}{2a}$. Use this information to evaluate each of the answer choices.

- Since the given parabola opens downward, the coefficient of the x^2-term must be less than 0. Choices (3) and (4) can therefore be eliminated. Determine which of the remaining answer choices represents a parabola with a vertical line of symmetry at $x = 1$.

- For the equation in choice (2), $a = -1$ and $b = -2$, so an equation of the axis of symmetry for this parabola is $x = -\dfrac{b}{2a} = -\dfrac{-2}{2(-1)} = -1$. Since the vertical line of symmetry for the given parabola is $x = 1$, choice (2) can be eliminated.

- For the equation in choice (1), $a = -1$ and $b = 2$, so an equation of the axis of symmetry for this parabola is $x = -\dfrac{b}{2a} = -\dfrac{2}{2(-1)} = \dfrac{-2}{-2} = 1$. Since the given parabola has $x = 1$ as a vertical line of symmetry, the parabola is the graph of the equation in choice (1).

The correct choice is **(1)**.

18. The length of each side of a triangle must be less than the sum of the lengths of the other two sides and greater than their difference. Test each of the answer choices in turn:

- Choice (1): Since 5 is *not* less than 1 + 3, this choice can be eliminated.

- Choice (2): Since 2 is *not* greater than 3 − 1, this choice can be eliminated.

- Choice (3): Since 3 < 1 + 3 and 3 > 3 − 1, the length of the third side may be 3.

- Choice (4): Since 4 is *not* less than 1 + 3, this choice can be eliminated.

The correct choice is **(3)**.

19. It is given that a girl can ski down a hill five times as fast as she can climb up the same hill. In other words, she can ski down in one fifth of the climbing time. Hence, if the girl can climb up the hill in x minutes, then she can ski down the hill in $\frac{1}{5}x$ minutes. It is also given that the girl can climb up the hill and ski down in a total of 9 minutes. Thus:

$$x + \frac{1}{5}x = 9$$

$$\frac{6}{5}x = 9$$

$$\frac{5}{6}\left(\frac{6}{5}x\right) = \frac{5}{6}(9)$$

$$x = \frac{45}{6} = 7.5$$

The correct choice is **(4)**.

20. The given problem is to subtract $3x^2 - 2x + 1$ from $2x^2 + 7x + 5$.

Write the polynomial to be subtracted underneath the other polynomial, with like terms aligned in the same column:

$$2x^2 + 7x + 5$$
$$-\quad\underline{3x^2 - 2x + 1}$$

Change to an equivalent addition example by reversing the sign of each each term of the polynomial to be subtracted:

$$+\begin{array}{r} 2x^2 + 7x + 5 \\ -3x^2 + 2x - 1 \\ \hline -x^2 + 9x + 4 \end{array}$$

Add like terms in each column:

The correct choice is **(1)**.

PART II

21. The accompanying diagram of a section of the city of Tacoma is given. High Road, State Street, and Main Street are parallel and 5 miles apart. Ridge Road is perpendicular to the three parallel streets. In the diagram the points

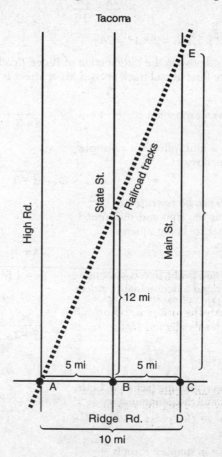

Tacoma

of intersection are labeled *A* through *E*. Since it is given that the distance between the intersection of Ridge Road and State Street and the point where the railroad tracks cross State Street is 12 miles, let $BC = 12$.

Use similar triangles to find the distance *DE* between the intersection of Ridge Road and Main Street and the point where the railroad tracks cross Main Street. Right $\triangle ABC$ is similar to right $\triangle ADE$. Since the lengths of corresponding sides of similar triangles are in proportion, set up a proportion and solve for *DE*:

$$\frac{BC}{AB} = \frac{DE}{AE}$$

$$\frac{12}{5} = \frac{DE}{5+5}$$

$$\frac{12}{5} = \frac{DE}{10}$$

Since $10 = \underline{2} \times 5$, $DE = \underline{2} \times 12 = 24$.

The distance between the intersection of Ridge Road and Main Street and the point where the railroad track crosses Main Street is **24** miles.

22. The given expression is:

$$\frac{x}{x+3} \div \frac{3x}{x^2-9}$$

Change to a multiplication example by inverting the divisor:

$$\frac{x}{x+3} \cdot \frac{x^2-9}{3x}$$

Since $x^2 - 9$ can be rewritten as $(x)^2 - (3)^2$, factor it as the sum and difference of the terms that are being squared:

$$\frac{x}{x+3} \cdot \frac{(x+3)(x-3)}{3x}$$

Divide out any factor that is common to a numerator and a denominator, since the quotient is 1:

$$\frac{\cancel{x}}{\cancel{x+3}} \cdot \frac{(\cancel{x+3})(x-3)}{3\cancel{x}}$$

Multiply the remaining factors in both the numerator and the denominator:

$$\frac{x-3}{3}$$

The quotient in simplest form is $\dfrac{x-3}{3}$.

23. It is given that Kerry is planning a rectangular garden with dimensions of 4 feet by 6 feet and that she wants half of the garden to be roses. According to Kerry, the rose plot will have dimensions of 2 feet by 3 feet. To determine if she is correct, reason as follows:

- The area of the rectangular garden is $4\,\text{ft} \times 6\,\text{ft} = 24\,\text{ft}^2$.

- The area of the rose plot is $2\,\text{ft} \times 3\,\text{ft} = 6\,\text{ft}^2$.

- Since 6 is *not* one half of 24, Kerry is not correct.

Kerry is incorrect.

24. It is given that the sum of the ages of the three Romano brothers is 63. If their ages can be represented as consecutive integers, then let x, $x + 1$, and $x + 2$ represent the ages of the three brothers. Thus:

$$x + (x + 1) + (x + 2) = 63$$
$$3x + 3 = 63$$
$$3x = 63 - 3$$
$$x = \frac{60}{3} = 20$$

Hence, the age of the youngest brother, x, is 20. The age of the middle brother, $x + 1$, is $20 + 1 = 21$. The middle brother is **21** years old.

25. It is given that Alan, Becky, Jesus, and Mariah are members of the chess club. If two of these students will be selected to represent the school at a national convention, then the number of possible combinations of two students can be determined using one of two methods.

<u>Method 1: Make an Organized List</u>
List pairs of students without regard to order:

Alan, Becky	Becky, Jesus
Alan, Jesus	Becky, Mariah
Alan, Mariah	Jesus, Mariah

Hence, **6** different combinations of two students are possible.

Method 2: Use the Combinations Formula

Use $_nC_r$ for selecting r objects from a set of n objects without regard to order. Evaluate $_nC_r$ for $n = 4$ and $r = 2$ by using a scientific calculator or by substituting in the formula

$$_nC_r = \frac{n!}{r!(n-r)!}$$

Let $n = 4$ and $r = 2$:

$$= \frac{4!}{2!(4-2)!}$$

$$= \frac{4 \times 3 \times \cancel{2 \times 1}}{\cancel{2 \times 1} \times (2 \times 1)}$$

$$= \frac{4 \times 3}{2 \times 1}$$

$$= \frac{12}{2}$$

$$= 6$$

Hence, **6** different combinations of two students are possible.

PART III

26. It is given that John, Dan, Karen, and Beth went to a costume ball as Anthony and Cleopatra, and Romeo and Juliet. Pick out the facts stated in the problem:

Fact 1: John got the costumes for Romeo and Cleopatra, but not his own.

Fact 2: Dan saw the costumes for Juliet and himself.

Fact 3: Karen went as Anthony.

Fact 4: Beth drove two of her friends, who were dressed as Anthony and Cleopatra.

Two methods can be used to determine the costume John wore.

Method 1: Use Logical Reasoning

Since John got the costumes for Romeo and Cleopatra, but not his own (fact 1), John was dressed as either Anthony or Juliet. Karen went as Anthony (fact 3), so John went as Juliet.

John wore the costume for **Juliet**.

Method 2: Make a Table

Organize the given information as shown in the accompanying table. Label the horizontal rows with the names of the partygoers and the vertical columns with the names of the costume characters. Then record each given fact in the table by entering a "No" or "Yes" depending on whether the person named in the leftmost column wore the costume in the table heading. Complete the table by entering "Yes" in any column or row that already has three entries of "No," and use "No" to fill in any column or row that already contains a "Yes."

	Anthony	**Cleopatra**	**Romeo**	**Juliet**
John	No	No	No	Yes
Dan	No	Yes	No	No
Karen	Yes	No	No	No
Beth	No	No	Yes	No

The intersection of a row and a column shows if the costume was worn by a partygoer.

As shown in the first row of the table, John wore the costume for **Juliet**.

27. It is given that a worker measures the length of a hiking trail using a device with a 2-foot-diameter wheel that counts the number of revolutions the wheel makes. The length of the trail can be determined by multiplying the circumference of the wheel by the number of revolutions counted at the end of the trail.

- The circumference, C, of the wheel is π times the 2-foot diameter. Hence, $C = \pi \times 2 \approx 3.14 \times 2 = 6.28 \, \text{ft}$.

- Since it is given that the device reads 1,100.5 revolutions at the end of the trail, the length of the trail is $1,100.5 \times 6.28 \, \text{ft} = 6,911.14 \, \text{ft}$.

- One mile is 5,280 feet. Hence, the length of the trail is $\dfrac{6,911.14}{5,280} \approx 1.3089$ miles.

The length of the trail, correct to the *nearest tenth of a mile*, is **1.3** miles.

28. It is given that the coordinates of the endpoints of $\overline{AB}$ are $A(2,6)$ and $B(4,2)$. Using this segment, you are asked to determine if the image $\overline{A''B''}$ is the same regardless of the order in which a reflection in the x-axis and a dilation are performed. Two methods may be used.

<u>Method 1: Work with Coordinates</u>

To reflect a point in the x-axis, replace the y-coordinate with its opposite. Hence, the image of the reflection of $\overline{AB}$ in the x-axis is $\overline{A'B'}$ with endpoints $A'(2,-6)$ and $B'(4,-2)$. To dilate a segment, multiply the coordinates of its endpoints by the scale factor. Hence, the image of $\overline{A'B'}$ after it is dilated by $\dfrac{1}{2}$ is $\overline{A''B''}$ with endpoints $A''\left(\dfrac{2}{2}, \dfrac{-6}{2}\right) = A''(1,-3)$ and $B''\left(\dfrac{4}{2}, \dfrac{-2}{2}\right) = B''(2,-1)$.

Now reverse the order of the transformations. After $\overline{AB}$ is dilated by $\frac{1}{2}$, its image is $\overline{A'B'}$ with endpoints $A'\left(\frac{2}{2},\frac{6}{2}\right) = A'(1,3)$ and $B'\left(\frac{4}{2},\frac{2}{2}\right) = B'(2,1)$.

The image of $\overline{A'B'}$ after it is reflected in the x-axis is $\overline{A''B''}$ with endpoints $A''(1,-3)$ and $B''(2,-1)$.

Since the coordinates of the endpoints of $\overline{A''B''}$ do not change, the image $\overline{A''B''}$ is the same regardless of the order of the two transformations. The correct answer is **yes**.

Method 2: Use Graph Paper

Graph both sequences of transformations on the same set of axes and show that the final images exactly coincide, as shown in the accompanying diagram.

Since the image $\overline{A''B''}$ is the same regardless of the transformation sequence, the correct answer is **yes**.

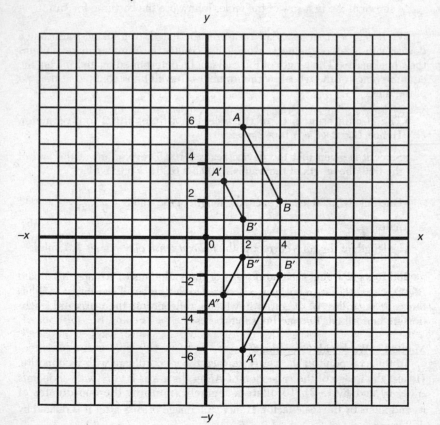

29. The fraction of homes in Glacier County that had power on Wednesday can be found by one of two methods.

Method 1: Use Arithmetic Reasoning

It is given that the three headlines quoted below were reported in the *Glacier County Times* on Monday, Tuesday, and Wednesday. To determine the portion of homes that had power on Wednesday, find the fractional loss or restoration of power each day.

- *Monday*: "Ice Storm Devastates County—8 out of every 10 homes lose electrical power." Hence, $\frac{8}{10}$ of homes lost electrical power, leaving $\frac{2}{10}$ of homes with power.

- *Tuesday*: "Restoration Begins—Power restored to $\frac{1}{2}$ of affected homes." Hence, $\frac{1}{2} \times \frac{8}{10}$ or $\frac{4}{10}$ of homes had power restored, meaning that $\frac{2}{10} + \frac{4}{10} = \frac{6}{10}$ of the homes had power on Tuesday.

- *Wednesday*: "More Freezing Rain—Power lost by 20% of homes that had power on Tuesday." Hence, 80% of the homes that had power on Tuesday now have power. Since 80% of $\frac{6}{10}$ is

$$\frac{80}{100} \times \frac{6}{10} = \frac{480}{1,000} = \frac{48}{100}$$

the portion of homes that had electrical power on Wednesday is $\frac{48}{100}$ (or any equivalent fraction) or **0.48** or **48%**.

Method 2: Use Particular Numbers in a Table

Suppose there are 100 homes. As you read the problem, organize the information in tabular form, as shown in the accompanying table. Each day the sum of the number of homes without power and the number of homes with power must be equal to 100. Thus, column (3) can be completed by subtracting the number in column (2) from 100. For example, since $\frac{8}{10} = \frac{80}{100}$, 80 homes lost power on Monday while 100 − 80 = 20 homes had power on Monday.

(1)	(2)	(3)
Day	Homes without Power	Homes with Power
Monday	80	$100 - 80 = 20$
Tuesday	$\frac{1}{2} \times 80 = 40$	$100 - 40 = 60$
Wednesday	$40 + (0.20 \times 60) = 40 + 12 = 52$	$100 - 52 = 48$

Hence, the portion of homes that had electrical power on Wednesday is $\frac{48}{100}$ (or any equivalent fraction) or **0.48** or **48%**.

30. It is given that Katrina hikes 5 miles north, 7 miles east, and then 3 miles north again, as shown in the accompanying map. The four points on her journey are labeled A through D.

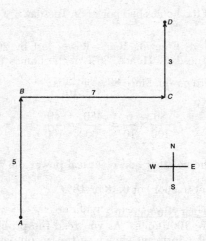

- To determine how far, in a straight line, Katrina is from her starting point, A, draw $\overline{AD}$. Then complete rectangle $BCDE$, as shown in the accompanying diagram.

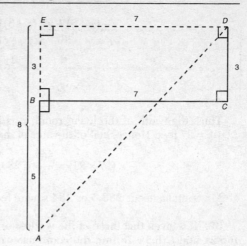

- Since opposite sides of a rectangle have the same length, $ED = BC = 7$ and $BE = CD = 3$. Thus, $AE = 5 + 3 = 8$.

- Since AED is a right triangle, use the Pythagorean theorem to find AD:

$$(AD)^2 = (AF)^2 + (FD)^2$$
$$= (8)^2 + (7)^2$$
$$= 64 + 49$$
$$= 113$$
$$AD = \sqrt{113} \approx 10.63$$

Correct to the *nearest tenth of a mile*, Katrina is **10.6** miles from her starting point.

PART IV

31. It is given that Mr. Santana wants to carpet exactly half of his rectangular living room, whose perimeter is 96 feet. It is also given that the length of the room is 6 feet greater than the width. To find the number of square feet of carpeting needed, first find the dimensions of the living room.

If x is the width of the living room, then $x + 6$ is the length of the living room, as shown in the accompanying diagram. Since the perimeter of a rectangle is the sum of the lengths of its sides:

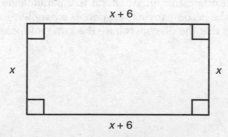

$$x + (x+6) + x + (x+6) = 96$$
$$4x + 12 = 96$$
$$4x = 96 - 12$$
$$x = \frac{84}{4} = 21$$

Thus, the width of the living room, x, is 21 feet and the length, $x + 6$, is $21 + 6 = 27$ feet. Hence, half of the area of the living room is

$$\frac{1}{2} \times (27 \times 21) = \frac{567}{2} = 283.5 \text{ square feet}$$

Mr. Santana needs **283.5** or **284** square feet of carpeting.

32. It is given that three of the vertices of parallelogram $ABCD$ are $A(0,0)$, $B(5,2)$, and $C(6,5)$. To find the coordinates of point $D(x,y)$, use the fact that the diagonals of a parallelogram must have the same midpoint.

• The coordinates of the midpoint of diagonal $\overline{AC}$ are the averages of the x- and y-coordinates of the endpoints of $\overline{AC}$. Hence, the midpoint of $\overline{AC}$ is

$$\left(\frac{0+6}{2}, \frac{0+5}{2} \right) = \left(\frac{6}{2}, \frac{5}{2} \right)$$

• Since the diagonals of a parallelogram have the same midpoint, the midpoint of diagonal $\overline{DB}$ is also $\left(\frac{6}{2}, \frac{5}{2} \right)$. Thus:

$$\left(\frac{x+5}{2}, \frac{y+2}{2} \right) = \left(\frac{6}{2}, \frac{5}{2} \right)$$

This means that $x + 5 = 6$, so $x + 1$; and $y + 2 = 5$, so $x = 3$.

• The coordinates are **D(1,3)**.

To justify that the figure is a parallelogram, plot vertices A, B, C, and D on the given set of axes and sketch the quadrilateral. If the diagonals of a quadrilateral bisect each other, the quadrilateral is a parallelogram. Quadrilateral $ABCD$, shown in the accompanying diagram, is a parallelogram, since $D(x,y)$ was determined so that the diagonals have the same midpoint and, as a result, bisect each other.

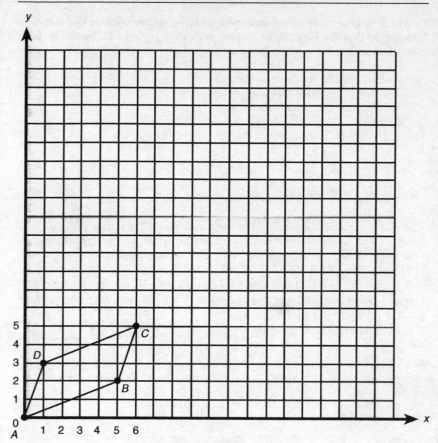

33. It is given that a 10-foot ladder is to be placed against the side of a building so that the base of the ladder makes an angle of 72° with the level ground, as shown in the accompanying diagram.

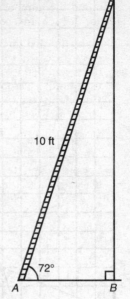

- To find how far the base of the ladder should be from the side of the building, use the cosine ratio in right $\triangle ABC$:

$$\cos 72° = \frac{AB}{AC}$$

$$0.3090 = \frac{AB}{10\,\text{ft}}$$

$$10\,\text{ft} \times 0.3090 = AB$$

$$3.09\,\text{ft} = AB$$

Hence, $AB = 3.09\,\text{ft} \times 12\dfrac{\text{in}}{\text{ft}} = 37.08\,\text{in.}$

The base of the ladder, *correct to the nearest inch*, should be **37 inches**, or **3 feet 1 inch**, from the side of the building.

- To find how far up the side of the building the ladder will reach, use the sine ratio in right $\triangle ABC$:

$$\sin 72° = \frac{BC}{AC}$$

$$0.9511 = \frac{BC}{10\,\text{ft}}$$

$$10\,\text{ft} \times 0.9511 = BC$$

$$9.511\,\text{ft} = BC$$

Hence, $BC = 9.511\,\text{ft} \times 12\dfrac{\text{in}}{\text{ft}} = 114.13\,\text{in.}$

The distance the ladder will reach up the side of the building, *correct to the nearest inch*, is **114 inches**, or **9 feet 6 inches**.

34. It is given that the telephone company will introduce a new area code and that each seven-digit telephone number for the new area code must meet the following conditions:

- The first digit cannot be 0 or 1.

- The first three digits cannot be 911 or 411.

To find the number of seven-digit telephone numbers that will be generated for the new area code given these two conditions, reason as follows:

- Since the first digit of a telephone number cannot be 0 or 1, any of the eight digits from 2 through 9 can be selected to fill the first position of the seven-digit telephone number. Any of the 10 digits from 0 to 9 can be used to fill each of the remaining six positions of the seven-digit number. Under the counting principle, the number of different seven-digit telephone numbers that do not begin with a 0 or a 1 is

$$\boxed{8} \times \boxed{10} \times \boxed{10} \times \boxed{10} \times \boxed{10} \times \boxed{10} \times \boxed{10} = 8 \times 10^6$$

- Since the 8×10^6 different telephone numbers include those numbers that begin with 911 or 411, we need to subtract from 8×10^6 the number of seven-digit telephone numbers that begin with 911 or 411. According to the counting principle, the number of seven-digit telephone numbers that begin with 911 is

first 3 digits are 911

$$\boxed{1} \times \boxed{1} \times \boxed{1} \times \boxed{10} \times \boxed{10} \times \boxed{10} \times \boxed{10} = 1 \times 10^4$$

Similarly, the number of seven-digit telephone numbers that begin with 411 is also 1×10^4.

- To find the number of seven-digit telephone numbers that meet the given conditions, subtract all the phone numbers beginning with 911 or 411 from the 8×10^6 possible numbers:

$$8 \times 10^6 - 1 \times 10^4 - 1 \times 10^4 = 800 \times 10^4 - 2 \times 10^4$$
$$= 798 \times 10^4$$

Hence, **798×10^4** or **7,980,000** new telephone numbers can be generated that meet the given conditions.

35. It is given that the width of Jack's rectangular dog pen is 2 yards less than the length and that the area of the pen is 15 square yards.

To figure out how many yards of fencing Jack needs to completely enclose the dog pen, first determine its dimensions by one of the two methods. Then use these dimensions to calculate the perimeter of the dog pen.

- Find the dimensions of the dog pen.

Method I: Use Trial and Error
Think of two positive integers that differ by 2 and whose product is 15.

Length	×	Width =	Area
6		4	24 ← Too high
5		3	15 ← This works!

<u>Method II: Use Algebra</u>

If the length of the dog pen is x, then the width of the pen is $x - 2$, as shown in the accompanying diagram.

Since area = length × width:

$15 = x(x - 2)$

$15 = x^2 - 2x$

Rewrite the quadratic equation so that all nonzero terms are on the same side:

$x - 2$　　Area = 15

x

$0 = x^2 - 2x - 15$ or, equivalently,

$$x^2 - 2x - 15 = 0$$

Factor the left side of the equation as the product of two binomials:

$$(x + ?)(x + ?) = 0$$

The missing terms of the binomial factors are the two integers whose sum is -2, the coefficient of the x-term in $x^2 - 2x - 15$, and whose product is 15, the constant term in $x^2 - 2x - 15$. Since $(-5) + (+3) = -2$ and $(-5)(+3) = -15$, the missing terms are -5 and $+3$:

$$(x - 5)(x + 3) = 0$$

If the product of two terms is 0, then at least one of these terms is 0:

$x - 5 = 0$　or　$x + 3 = 0$

Solve each first-degree equation:

$x = 5$　or　$x = -3$ Reject, since length cannot be negative.

Hence, the length of the dog pen is 5 yards and the width is $5 - 2 = 3$ yards.

• Using the dimensions found by either method above, calculate the perimeter of the dog pen. Since the perimeter of a rectangle is the sum of the lengths of its sides:

$$\text{Perimeter} = 5 + 3 + 5 + 3 = 16 \text{ yards}$$

Jack will need **16** yards of fencing to completely enclose the pen.

Topic	Question Numbers	Number of Points	Your Points	Your Percentage
1. Numbers; Properties of Real Numbers; Percent	10	2		
2. Operations on Rat'l. Numbers & Monomials	1	2		
3. Laws of Exponents for Integer Exponents; Scientific Notation	4	2		
4. Operations on Polynomials	20	2		
5. Square Root; Operations with Radicals	16	2		
6. Evaluating Formulas & Algebraic Expressions	—	—		
7. Solving Linear Eqs. & Inequalities	15	2		
8. Solving Literal Eqs. & Formulas for a Particular Letter	—	—		
9. Alg. Operations (including factoring)	6, 22	$2 + 2 = 4$		
10. Solving Quadratic Eqs.; graphs of parabolas & circles	12, 17	$2 + 2 = 4$		
11. Coordinate Geometry (Graphs of linear eqs; slope; midpoint; distance)	9, 32	$2 + 4 = 6$		
12. Systems of Linear Eqs. & Inequalities (alg. & graph. solutions)	13	2		
13. Mathematical Modeling using: Eqs.; Tables; Graphs of Linear Eqs.; Direct Variation	5	2		
14. Linear-Quad systems (alg. & graph. solutions).	—	—		
15. Word Problems Requiring Arith. or Alg. Reasoning	2, 19, 24, 29	$2 + 2 + 2 + 3 = 9$		
16. Areas, Perims., Circums., Vols. of Common Figures	7, 23, 27, 31, 35	$2 + 2 + 3 + 4 + 4 = 15$		
17. Angle & Line Relationships (suppl., compl., vertical angles; parallel lines; congruence)	—	—		
18. Ratio & Proportion (incl. similar polygons)	21	2		
19. Pythagorean Theorem	30	3		
20. Right Triangle Trig. & Indirect Measurement	33	4		
21. Logic (conditionals; logically equiv. statements; valid arguments)	14, 26	$2 + 3 = 5$		
22. Probability (incl. tree diagrams & sample spaces)	11	2		
23. Counting Methods & Sets	34	4		
24. Permutations & Combinations	25	2		
25. Statistics (mean, percentiles, quartiles; freq. dist., histograms, stem & leaf plots; box-and-whisker plots; circle graphs	8	2		

Topic	Question Numbers	Number of Points	Your Points	Your Percentage
26. Properties of triangles & parallelograms	18	2		
27. Transformations (reflections, translations, rotations, dilations)	—	—		
28. Symmetry	—	—		
29. Area and transformations using Coordinates	28	3		
30. Locus & Constructions	3	2		
31. Dimensional Analysis & Units of Measurement	—	—		

MAP TO LEARNING STANDARDS

Key Ideas	Item Numbers
Mathematical Reasoning	14, 26
Number and Numeration	10, 27
Operations	1, 4, 16, 20, 22, 29
Modeling/Multiple Representation	2, 3, 6, 18, 21, 28, 32
Measurement	5, 7, 8, 19, 23, 30, 31, 33
Uncertainty	11, 25, 34
Patterns/Functions	9, 12, 13, 15, 17, 24, 35

HOW TO CONVERT YOUR RAW SCORE TO YOUR MATH A REGENTS EXAMINATION SCORE

Below is the conversion chart that must be used to determine your final score on the August 2000 Regents Examination in Math A. To find your final exam score, locate in the column labeled "Raw Score" the total number of points you scored out of a possible 85 points. Since partial credit is allowed in Parts II, III, and IV of the test, you may need to approximate the credit you would receive for a solution that is not completely correct. Then locate in the adjacent column to the right the scaled score that corresponds to your raw score. The scaled score is your final Math A Regents Examination score.

Raw Score	Scaled Score	Raw Score	Scaled Score	Raw Score	Scaled Score
85	100	56	82	27	48
84	99	55	81	26	47
83	99	54	80	25	46
82	99	53	79	24	44
81	99	52	78	23	43
80	98	51	76	22	42
79	98	50	75	21	41
78	98	49	74	20	40
77	97	48	73	19	39
76	97	47	72	18	38
75	97	46	71	17	37
74	96	45	70	16	36
73	95	44	68	15	35
72	95	43	67	14	34
71	94	42	66	13	33
70	94	41	65	12	32
69	93	40	64	11	30
68	92	39	63	10	29
67	91	38	61	9	28
66	91	37	60	8	27
65	90	36	59	7	26
64	89	35	58	6	25
63	88	34	57	5	24
62	87	33	55	4	22
61	87	32	54	3	18
60	86	31	53	2	13
59	85	30	52	1	7
58	84	29	50	0	0
57	83	28	49		

Examination
January 2001
Math A

PART I

Answer all questions in this part. Each correct answer will receive 2 credits. No partial credit will be allowed. Record your answers in the spaces provided. [40]

1 There are 461 students and 20 teachers taking buses on a trip to a museum. Each bus can seat a maximum of 52. What is the *least* number of buses needed for the trip?

(1) 8   (3) 10

(2) 9   (4) 11   1 _____

2 In right triangle ABC, $m\angle C = 3y - 10$, $m\angle B = y + 40$, and $m\angle A = 90$. What type of right triangle is triangle ABC?

(1) scalene   (3) equilateral

(2) isosceles   (4) obtuse   2 _____

3 If $x > 0$, the expression $\left(\sqrt{x}\right)\left(\sqrt{2x}\right)$ is equivalent to

(1) $\sqrt{2x}$ (3) $x^2\sqrt{2}$

(2) $2x$ (4) $x\sqrt{2}$ 3 _____

4 Three times as many robins as cardinals visited a bird feeder. If a total of 20 robins and cardinals visited the feeder, how many were robins?

(1) 5   (3) 15
(2) 10   (4) 20   4____

5 One of the factors of $4x^2 - 9$ is

(1) $(x + 3)$ (3) $(4x - 3)$
(2) $(2x + 3)$ (4) $(x - 3)$ 5____

6 At a school fair, the spinner represented in the accompanying diagram is spun twice.

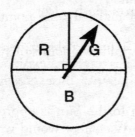

What is the probability that it will land in section G the first time and then in section B the second time?

(1) $\frac{1}{2}$ (3) $\frac{1}{8}$

(2) $\frac{1}{4}$ (4) $\frac{1}{16}$ 6____

7 If a and b are integers, which equation is always true?

(1) $\dfrac{a}{b} = \dfrac{b}{a}$ (3) $a - b = b - a$

(2) $a + 2b = b + 2a$ (4) $a + b = b + a$ 7____

8 The sum of $3x^2 + 4x - 2$ and $x^2 - 5x + 3$ is

   (1) $4x^2 + x - 1$ (3) $4x^2 + x + 1$

   (2) $4x^2 - x + 1$ (4) $4x^2 - x - 1$ 8 _____

9 If $x \uparrow 0$, the expression $\dfrac{x^2 + 2x}{x}$ is equivalent to

   (1) $x + 2$ (3) $3x$

   (2) 2 (4) 4 9 _____

10 Helen is using a capital H in an art design. The H has

   (1) only one line of symmetry
   (2) only two points of symmetry
   (3) two lines of symmetry and only one point of symmetry
   (4) two lines of symmetry and two points of symmetry   10 _____

11 The distance from Earth to the Sun is approximately 93 million miles. A scientist would write that number as

   (1) 9.3×10^6 (3) 93×10^7

   (2) 9.3×10^7 (4) 93×10^{10} 11 _____

12 Given the statement: "If two sides of a triangle are congruent, then the angles opposite these sides are congruent."

Given the converse of the statement: "If two angles of a triangle are congruent, then the sides opposite these angles are congruent."

What is true about this statement and its converse?

(1) Both the statement and its converse are true.
(2) Neither the statement nor its converse is true.
(3) The statement is true but its converse is false.
(4) The statement is false but its converse is true.   12 _____

13 Which equation could represent the relationship between the x and y values shown in the accompanying table?

x	y
0	2
1	3
2	6
3	11
4	18

(1) $y = x + 2$ (3) $y = x^2$
(2) $y = x^2 + 2$ (4) $y = 2^x$ 13 _____

14 A locker combination system uses three digits from 0 to 9. How many different three-digit combinations with no digit repeated are possible?

(1) 30   (3) 720
(2) 504   (4) 1,000   14 _____

15 What is the slope of line ℓ in the accompanying diagram?

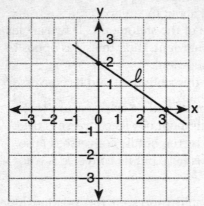

(1) $-\frac{3}{2}$ (3) $\frac{2}{3}$

(2) $-\frac{2}{3}$ (4) $\frac{3}{2}$ 15 _____

16 If $bx - 2 = K$, then x equals

(1) $\frac{K}{b} + 2$ (3) $\frac{2 - K}{b}$

(2) $\frac{K - 2}{b}$ (4) $\frac{K + 2}{b}$ 16 _____

17 In a molecule of water, there are two atoms of hydrogen and one atom of oxygen. How many atoms of hydrogen are in 28 molecules of water?

(1) 14   (3) 42

(2) 29   (4) 56   17 _____

18 From January 3 to January 7, Buffalo recorded the following daily high temperatures: 5°, 7°, 6°, 5°, and 7°. Which statement about the temperatures is true?

(1) mean = median   (3) median = mode
(2) mean = mode   (4) mean < median   18 ____

19 In which of the accompanying figures are segments *XY* and *YZ* perpendicular?

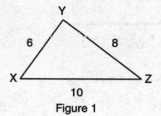 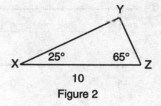

Figure 1   Figure 2

(1) figure 1, only
(2) figure 2, only
(3) both figure 1 and figure 2
(4) neither figure 1 nor figure 2   19 ____

20 Let x and y be numbers such that $0 < x < y < 1$, and let $d = x - y$. Which graph could represent the location of d on the number line?

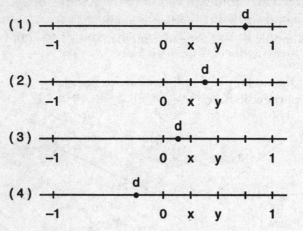

20 _____

PART II

Answer all questions in this part. Each correct answer will receive 2 credits. Clearly indicate the necessary steps, including appropriate formula substitutions, diagrams, graphs, charts, etc. For all questions in this part, a correct numerical answer with no work shown will receive only 1 credit. [10]

21 The accompanying graph shows Marie's distance from home (A) to work (F) at various times during her drive.

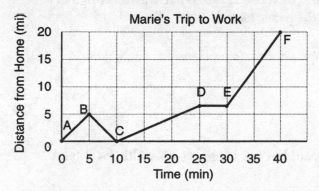

a Marie left her briefcase at home and had to return to get it. State which point represents when she turned back around to go home and explain how you arrived at that conclusion.

b Marie also had to wait at the railroad tracks for a train to pass. How long did she wait?

22 Sue bought a picnic table on sale for 50% off the original price. The store charged her 10% tax and her final cost was $22.00. What was the original price of the picnic table?

23 A cardboard box has length $x - 2$, width $x + 1$, and height $2x$.

a Write an expression, in terms of x, to represent the volume of the box.

b If $x = 8$ centimeters, what is the number of cubic centimeters in the volume of the box?

24 The coordinates of the endpoints of $\overline{AB}$ are $A(0,2)$
    and $B(4,6)$. Graph and state the coordinates of A'
    and B', the images of A and B after $\overline{AB}$ is reflected
    in the x-axis.

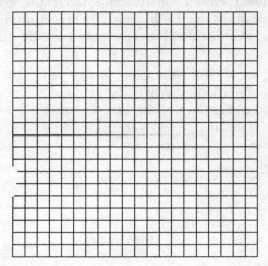

25 Two trains leave the same station at the same time
    and travel in opposite directions. One train travels
    at 80 kilometers per hour and the other at 100 kilo-
    meters per hour. In how many hours will they be
    900 kilometers apart?

PART III

Answer all questions in this part. Each correct answer will receive 3 credits. Clearly indicate the necessary steps, including appropriate formula substitutions, diagrams, graphs, charts, etc. For all questions in this part, a correct numerical answer with no work shown will receive only 1 credit. [15]

26 Sal has a small bag of candy containing three green candies and two red candies. While waiting for the bus, he ate two candies out of the bag, one after another, without looking. What is the probability that both candies were the same color?

27 Steve has a treasure map, represented in the accompanying diagram, that shows two trees 8 feet apart and a straight fence connecting them. The map states that treasure is buried 3 feet from the fence and equidistant from the two trees.

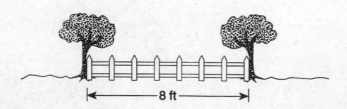

a Sketch a diagram to show all the places where the treasure could be buried. Clearly indicate in your diagram where the treasure could be buried.

b What is the distance between the treasure and one of the trees?

28 In the accompanying figure, two lines intersect, $m\angle 3 = 6t + 30$, and $m\angle 2 = 8t - 60$. Find the number of degrees in $m\angle 1$.

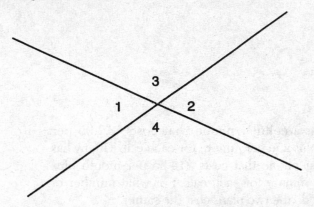

29 Mark says, "The number I see is odd." Jan says, "That same number is prime." The teacher says, "Mark is correct or Jan is correct." Some integers would make the teacher's statement true while other integers would make it false. Give and explain *one* example of when the teacher's statement is true. Give and explain *one* example of when the teacher's statement is false.

30 Juan has a cellular phone that costs $12.95 per month plus 25¢ per minute for each call. Tiffany has a cellular phone that costs $14.95 per month plus 15¢ per minute for each call. For what number of minutes do the two plans cost the same?

PART IV

Answer all questions in the part. Each correct answer will receive 4 credits. Clearly indicate the necessary steps, including appropriate formula substitutions, diagrams, graphs, charts, etc. For all questions in this part, a correct numerical answer with no work shown will receive only 1 credit. [20]

31 Solve algebraically for x: $\dfrac{1}{x} = \dfrac{x+1}{6}$

32 On a science quiz, 20 students received the following scores: 100, 95, 95, 90, 85, 85, 85, 80, 80, 80, 80, 75, 75, 75, 70, 70, 65, 65, 60, 55.

Construct a statistical graph, such as a histogram or a stem-and-leaf plot, to display this data. [*Be sure to title the graph and label all axes or parts used.*]

If your type of plot requires a grid, show your work here.

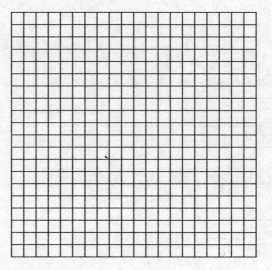

If no grid is necessary, show your work here.

33 John uses the equation $x^2 + y^2 = 9$ to represent the shape of a garden on graph paper.

a Graph $x^2 + y^2 = 9$ on the accompanying grid.

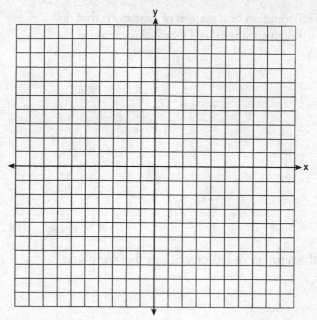

b What is the area of the garden to the *nearest square unit*?

34 There were 100 more balcony tickets than main-floor
    tickets sold for a concert. The balcony tickets sold for
    $4 and the main-floor tickets sold for $12. The total
    amount of sales for both types of tickets was $3,056.

a Write an equation or a system of equations that de-
     scribes the given situation. Define the variables.

b Find the number of balcony tickets that were sold.

35 Find, to the *nearest tenth of a foot*, the height of the
   tree represented in the accompanying diagram.

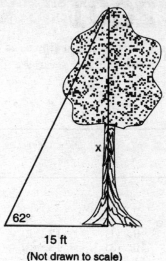

15 ft

(Not drawn to scale)

Answers
January 2001
Math A

Answer Key

PART I

1. (3)	**5.** (2)	**9.** (1)	**13.** (2)	**17.** (4)
2. (1)	**6.** (3)	**10.** (3)	**14.** (3)	**18.** (1)
3. (4)	**7.** (4)	**11.** (2)	**15.** (2)	**19.** (3)
4. (3)	**8.** (2)	**12.** (1)	**16.** (4)	**20.** (4)

PART II

21. a. B **b.** 5 minutes

22. \$40.00

23. a. $(x-2)(x+1)(2x)$ or any equivalent expression such as $V = 2x^3 - 2x^2 - 4x$

  b. 864

24. $A'(0,-2)$ and $B(4,-6)$

25. 5

PART III

26. $\dfrac{8}{20}$

27. a. Two possible locations
  b. 5 feet

28. 120

29. See *Answers Explained* section.

30. 20

PART IV

31. 2 and −3
32. See *Answers Explained* section.
33. **a.** See *Answers Explained* section.
    b. 28

34. **a.** See *Answers Explained* section.
    b. 266
35. 28.2

Parts II–IV You are required to show how you arrived at your answers. For sample methods of solution, see Answers Explained section.

Answers Explained

PART I

1. It is given that there are 461 students and 20 teachers taking buses and that each bus can seat a maximum of 52 people.
- The total number of passengers is 461 + 20 = 481.
- Hence, 481 ÷ 52 = 9.25 buses are needed.
- Since a fractional part of a bus is not possible, 9.25 must be rounded up to 10.

The correct choice is **(3)**.

2. It is given that in right triangle ABC, m$\angle C$ = 3y − 10, m$\angle B$ = y + 40, and m$\angle A$ = 90.
- A right triangle cannot be equilateral since the hypotenuse must be longer than either of the other two sides. Hence, eliminate choice (3).
- If a right triangle contained an obtuse angle in addition to the right angle, the sum of the measures of the three angles of the triangle would be greater than 180, which is not possible. A right triangle cannot be obtuse. Hence, eliminate choice (4).
- To determine if the right triangle is scalene or isosceles, find the measures of angles B and C. Since the sum of the measures of the acute angles of a right triangle is 90,

$$m\angle B + m\angle C = 90$$
$$(y + 40) + (3y − 10) = 90$$
$$4y + 30 = 90$$
$$4y = 60$$
$$y = \frac{60}{4} = 15$$

Thus, m$\angle B$ = y + 40 = 15 + 40 = 55 and m$\angle C$ = 90 − 55 = 35. The acute angles of the right triangle are not congruent and, as a result, the sides of the right triangle opposite these angles are not congruent. Since right triangle ABC has three sides with unequal lengths, it is a scalene triangle.

The correct choice is **(1)**.

3. If $x > 0$, then

$$\left(\sqrt{x}\right)\left(\sqrt{2x}\right) = \sqrt{2x^2} = \sqrt{x^2} \cdot \sqrt{2} = x\sqrt{2}$$

The correct choice is **(4)**.

4. It is given that three times as many robins as cardinals visited a bird feeder. If x represents the number of cardinals, then $3x$ represents the number of robins. Since it is also given that a total of 20 robins and cardinals visited the feeder,

$$x + 3x = 20$$
$$4x = 20$$
$$x = \frac{20}{4} = 5$$

The number of robins = $3x = 3(5) = 15$.

The correct choice is **(3)**.

5. Since the given expression, $4x^2 - 9$, represents the difference between two squares, it can be factored as the sum and difference of the two terms that are being squared:

$$4x^2 - 9 = (2x)^2 - (3)^2$$
$$= (2x + 3)(2x - 3)$$

Hence, one of the factors of $4x^2 - 9$ is $x + 3$.

The correct choice is **(2)**.

6. It is given that at a school fair, the spinner in the accompanying diagram is spun twice. Since regions R and G are each one quarter of the circle and region B is one half of circle,

$$P(\text{spinner lands in R}) = \frac{1}{4}$$
$$P(\text{spinner lands in G}) = \frac{1}{4}$$
$$P(\text{spinner lands in B}) = \frac{1}{2}$$

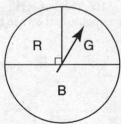

To find the probability that the spinner will land in section G the first time and then in section B the second time, multiply together the probabilities that each event will occur:

$$P(\text{lands in G, B}) = P(\text{lands in G}) \times P(\text{lands in B})$$
$$= \quad \frac{1}{4} \quad \times \quad \frac{1}{2}$$
$$= \frac{1}{8}$$

The correct choice is **(3)**.

7. If a and b are integers, then the order in which these numbers are divided or subtracted matters, and so choices (1) and (3) can be eliminated. Choice (2) states that $a + 2b = b + 2a$, which is true only when $a = b$, a fact that cannot be assumed. Since the order in which two integers are added does not matter, the equation $a + b = b + a$ is always true.

The correct choice is **(4)**.

8. To find the sum of $3x^2 + 4x - 2$ and $x^2 - 5x + 3$, write the second polynomial underneath the first. Then add like terms:

$$\begin{array}{r} 3x^2 + 4x - 2 \\ +\ \ x^2 - 5x + 3 \\ \hline 4x^2 -\ \ x + 1 \end{array}$$

The correct choice is **(2)**.

9. If $x \uparrow 0$, then the given expression $\dfrac{x^2 + 2x}{x}$ can be written in an equivalent form by factoring the numerator and simplifying:

$$\frac{x^2 + 2x}{x} = \frac{\overset{1}{\cancel{x}}(x + 2)}{\cancel{x}} = x + 2$$

The correct choice is **(1)**.

10. The letter **H** has a line of vertical symmetry and a line of horizontal symmetry, as shown in the accompanying diagram:

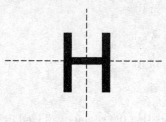

If a figure coincides with its image after a 180° rotation about a point, then the figure has a point of symmetry. The letter **H**, therefore, has a point of symmetry since it coincides with its image after a 180° rotation about the point at which the horizontal and vertical lines of symmetry intersect.

The correct choice is **(3)**.

11. A scientist would write 93 million miles in *scientific notation*. A number is in scientific notation when it is written as the product of a number between 1 and 10, and a power of 10. Since one million can be written as $1,000,000 = 10^6$,

$$93 \text{ million} = 93 \times 10^6$$
$$= (9.3 \times 10^1) \times 10^6$$
$$= 9.3 \times 10^7.$$

The correct choice is **(2)**.

12. The given statement, "If two sides of a triangle are congruent, then the angles opposite these sides are congruent," is true. The converse of this statement, "If two angles of a triangle are congruent, then the sides opposite these angles are congruent," is also true. Hence, both the statement and its converse are true.

The correct choice is **(1)**.

13. To determine which equation could represent the relationship between the values of x and y shown in the accompanying table, test each ordered pair in each equation until you find the equation that works for all the ordered pairs in the table.

- Choice (1): $y = x + 2$. This equation does not work for $(3,11)$ since $y = 3 + 2 \uparrow 11$.
- Choice (2): $y = x^2 + 2$. This equation works for each ordered pair:
  For $(0,2)$, $y = 0^2 + 2 = 2$.
  For $(1,3)$, $y = 1^2 + 2 = 1 + 2 = 3$.
  For $(2,6)$, $y = 2^2 + 2 = 4 + 2 = 6$.
  For $(3,11)$, $y = 3^2 + 2 = 9 + 2 = 11$.
  For $(4,18)$, $y = 4^2 + 2 = 16 + 2 = 18$.
- Choice (3): $y = x^2$. This equation does not work for $(0,2)$ since $y = 0^2 = 0 \uparrow 2$.
- Choice (4): $y = 2^x$. This equation does not work for $(0,2)$ since $y = 2^0 = 1 \uparrow 2$.

x	y
0	2
1	3
2	6
3	11
4	18

The correct answer is choice **(2)**.

14. It is given that a locker combination system uses three digits from 0 to 9. To find the number of different three-digit lock combinations with no digit repeated that are possible, multiply together the number of digits that can be used to fill each of the three positions in the lock combination:

- Since there are 10 digits from 0 to 9, any of the 10 digits can be used as the first digit of the lock combination:

$$\boxed{10} \times \square \times \square$$

- The second digit of the lock combination can be selected from any of the remaining 9 digits:

$$\boxed{10} \times \boxed{9} \times \square$$

- Any of the remaining 8 digits can be used as the last digit of the lock combination:

$$\boxed{10} \times \boxed{9} \times \boxed{8} = 720$$

The correct choice is **(3)**.

15. To find the slope of line ℓ in the accompanying diagram, pick any two points on the line and calculate the change in the y-coordinates over the change in the corresponding x-coordinates. Using $A(0,2)$ and $B(3,0)$,

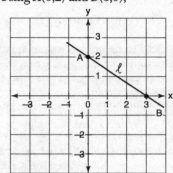

- Change in $y = 0 - 2 = -2$.
- Change in $x = 3 - 0 = 3$.
- Slope $= \dfrac{\text{change in } y}{\text{change in } x} = \dfrac{-2}{3}$.

The correct choice is **(2)**.

16. If $bx - 2 = K$, then $bx = K + 2$, so $x = \dfrac{K + 2}{b}$.

The correct choice is **(4)**.

17. It is given that in one molecule of water, there are two atoms of hydrogen. Hence, in 28 molecules of water, there are $28 \times 2 = 56$ atoms of hydrogen.

The correct choice is **(4)**.

18. It is given that from January 3 to January 7, Buffalo recorded the following daily high temperatures: $5°$, $7°$, $6°$, $5°$, and $7°$.
- The mode is the score that occurs most often. Hence, the mode of the five temperatures is $5°$ *and* $7°$.
- The median is the middle score when the scores are arranged in size order. When arranged in size order, the five temperatures are $5°$, $5°$, $6°$, $7°$, and $7°$. Hence, the median is $6°$.
- The mean (average) of a set of scores is the score obtained by dividing the sum of the scores by the number of scores. Hence, the mean is

$$\frac{5° + 5° + 6° + 7° + 7°}{5} = \frac{30°}{5} = 6°.$$

Compare the mode, median, and mean to find that mean = median.

The correct choice is (**1**).

19. In the accompanying figures, segments XY and YZ are perpendicular if they form a right angle.

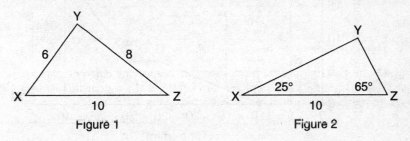

Figure 1   Figure 2

- According to the converse of the Pythagorean Theorem, if the square of the length of the longest side of a triangle is equal to the sum of the squares of the lengths of the other two sides, then the triangle is right triangle. Since $10^2 = 100$ *and* $6^2 + 8^2 = 36 + 64 = 100$, $10^2 = 6^2 + 8^2$, so Figure 1 is a right triangle in which Y is a right angle. This means segments XY and YZ are perpendicular.
- In Figure 2, $25° + 65° = 90°$, so the measure of the third angle of the triangle, angle Y, must be $180° - 90° = 90°$. Since angle Y is a right angle, segments XY and YZ are perpendicular.
- Hence, segments XY and YZ are perpendicular in both figures.

The correct choice is (**3**).

20. It is given that x and y represent numbers such that $0 < x < y < 1$. Since x and y are positive numbers and $x < y$, the difference $x - y$ must be negative. Since $d = x - y$, d is a negative number. The only graph in which d corresponds to a negative number is the graph in choice (4):

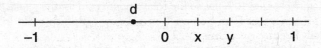

The correct choice is (**4**).

PART II

21.

Marie s Trip to Work

It is given that the accompanying graph shows Marie's distance from home (A) to work (F) at various times during her drive.

a. To find the point at which Marie turned around to go back home, look for the point after A at which the distance from home is 0. At point C Marie arrived back home. Hence, at the preceding point, B, Marie turned around to go back home.

Marie turned around to go back home at point **B**.

b. To find how long Marie waited for a train to pass, locate a horizontal segment. From D to E, Marie did not travel any distance. In this interval, the elapsed time was $25 - 20 = 5$ minutes.

Marie waited **5 minutes** for a train to pass.

22. It is given that Sue bought a picnic table on sale for 50% off the original price.
- If the original price was x, the sale price was $0.50x$.
- The store charged 10% tax, so the total amount paid for the picnic table was

$$\underbrace{0.50x}_{\text{sale price}} + \underbrace{(10\% \times 0.50x)}_{\text{tax on sale price}} = 0.50x + 0.05x = 0.55x$$

- Since the final cost was \$22.00, $0.55x = \$22.00$, $x = \dfrac{\$22.00}{0.55} = \$40.00$.

The original price of the picnic table was **\$40.00**.

23. It is given that a cardboard box has length $x - 2$, width $x + 1$, and height $2x$.

a. Since the volume of a box is length times width times height, the volume of the box is $(x - 2)(x + 1)(2x)$. After multiplying the factors together, the volume V of the box can also be represented by $V = 2x^3 - 2x^2 - 4x$.

b. If $x = 8$ centimeters, the number of cubic centimeters in the volume of the box is $(8 - 2)(8 + 1)(2 \cdot 8) = (6)(9)(16) = 864$.

The number of cubic centimeters in the volume of the box is **864**.

24. To reflect a point in the x-axis, replace the y-coordinate with its opposite, as shown in the accompanying figure.
- The reflection of $A(0,2)$ in the x-axis is $A'(0,-2)$.
- The reflection of $B(4,6)$ in the x-axis is $B'(4,-6)$.
- Connect $A'(0,-2)$ and $B'(4,-6)$ using a straightedge to obtain the image of $\overline{AB}$ after a reflection in the x-axis, as shown in the accompanying figure.

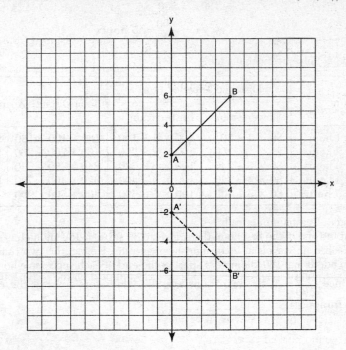

25. It is given that two trains leave the same station at the same time and travel in opposite directions. One train travels at 80 kilometers per hour, and the other at 100 kilometers per hour. To find the number of hours when the trains will be 900 kilometers apart, use one of the following three methods.

Method 1: Solve algebraically

Let x represent the number of hours the trains have traveled when they are 900 kilometers apart. Since rate multiplied by time equals distance,

- The slower train has traveled a distance of $80x$ kilometers.
- The faster train has traveled a distance of $100x$ kilometers.

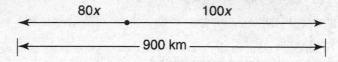

$$80x$$ $$100x$$

$$\overleftrightarrow{\hspace{3cm}\bullet\hspace{3cm}}$$

$$\vdash\!\!\!\!\!-\!\!\!-\!\!\!-\!\!\!-\!\!\! 900\text{ km} \!-\!\!\!-\!\!\!-\!\!\!-\!\!\dashv$$

- Since the sum of the distances traveled by the two trains is 900 kilometers,

$$80x + 100x = 900$$
$$180x = 900$$
$$x = \frac{900}{180} = \textbf{5 hours}$$

Method 2: Solve by making a table

Time	Distance Traveled (km)		Total Distance Apart (km)
	Slower Train	Faster Train	
After 1 hour	$80 \times 1 = 80$	$100 \times 1 = 100$	$80 + 100 = 180 \uparrow 900$
After 2 hours	$80 \times 2 = 160$	$100 \times 2 = 200$	$160 + 200 = 360 \uparrow 900$
After 3 hours	$80 \times 3 = 240$	$100 \times 3 = 300$	$240 + 300 = 540 \uparrow 900$
After 4 hours	$80 \times 4 = 320$	$100 \times 4 = 400$	$320 + 400 = 720 \uparrow 900$
After **5 hours**	$80 \times 5 = 400$	$100 \times 5 = 500$	$400 + 500 = 900$

Method 3: Reason numerically

- At the end of the first hour, the trains are $80 + 100 = 180$ kilometers apart.
- Each train travels at a constant average rate. Hence, the two trains are an additional 180 kilometers apart at the end of each successive hour.
- Since $\frac{900}{180} = 5$, the two trains are 900 kilometers apart at the end of **5 hours**.

PART III

26. It is given that, from a bag containing three green candies and two red candies, Sal eats two candies, one after another, without looking. The probability that both candies will be the same color is the sum of the probabilities that Sal eats two red candies or two green candies.

• Find P(red, red).

Since 2 of the 5 candies are red, the probability that the first candy eaten is red is $\frac{2}{5}$. Because 1 of the remaining 4 candies is red, the probability that the second candy eaten is red, given that the first candy eaten is red, is $\frac{1}{4}$. Hence, P(red, red) $= \frac{2}{5} \times \frac{1}{4} = \frac{2}{20}$.

• Find P(green, green).

Since 3 of the 5 candies are green, the probability that the first candy eaten is green is $\frac{3}{5}$. Because 2 of the remaining 4 candies are green, the probability that the second candy eaten is green, given that the first candy eaten is green, is $\frac{2}{4}$. Hence, P(green, green) $= \frac{3}{5} \times \frac{2}{4} = \frac{6}{20}$.

• Add the probabilities that both candies are red or both candies are green.

$$P(\text{same color candies}) = P(\text{red, red}) + P(\text{green, green})$$
$$= \frac{2}{20} + \frac{6}{20}$$
$$= \frac{8}{20}$$

The probability that both candies will be the same color is $\frac{8}{20}$.

27.

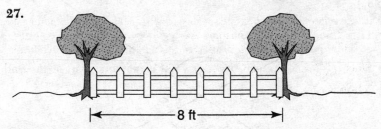

It is given that Steve has a treasure map showing two trees 8 feet apart and a straight fence connecting them, as shown in the accompanying diagram. The map states that treasure is buried 3 feet from the fence (locus condition 1) and equidistant from the two trees (locus condition 2).

a. Locate the treasure by finding the points at which the two locus conditions intersect:

- Since the treasure is buried 3 feet from the fence, the treasure is located along one of the segments parallel to the fence and at a distance of 3 feet from the fence, as shown in the accompanying diagram.

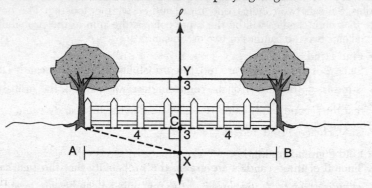

- Since the treasure is equidistant from the two trees, it is located along the perpendicular bisector ℓ of the segment AB joining the centers of the two trees, as shown in the accompanying diagram.
- The treasure could be buried at either point X or point Y where line ℓ intersects the parallel segments, as shown in the accompanying diagram.

b. If line ℓ intersects $\overline{AB}$ at point C, then $\triangle ACX$ is a right triangle with CX = 3 and $AC = \frac{1}{2} \times 8 = 4$. Thus, $\triangle ACX$ is a 3-4-5 right triangle in which the hypotenuse $AX =$ **5 feet**.

The distance between the treasure and one of the trees is **5 feet**.

28.

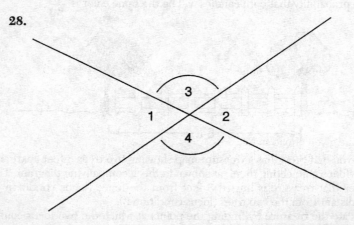

In the accompanying figure, $m\angle 3 = 6t + 30$ and $m\angle 2 = 8t - 60$.

- Since angles 2 and 3 form a linear pair, the sum of their measures is 180. Thus,

$$m\angle 2 + m\angle 3 = 180$$
$$(8t - 60) + (6t + 30) = 180$$
$$14t - 30 = 180$$
$$14t = 180 + 30$$
$$t = \frac{210}{14} = 15$$

- Angles 3 and 4 are vertical angles and, as a result, have the same measure.
- Hence, $m\angle 4 = m\angle 3 = 6(15) + 30 = 90 + 30 = 120$.

The number of degrees in $m\angle 4$ is **120**.

29. It is given that Mark says, "The number I see is odd," Jan says, "That same number is prime," and the teacher says, "Mark is correct or Jan is correct."

- The teacher's statement is true for any number that is odd or is prime, such as 2, 3, 5, 7, 9, 11, and so forth.
- The teacher's statement is false for any number that is even other than 2 because 2 is the only even number that is prime.

30. It is given that Juan has a cellular phone that costs \$12.95 per month plus 25¢ per minute for each call, and Tiffany has a cellular phone that costs \$14.95 per month plus 15¢ per minute for each call. Let x represent the number of minutes either cellular phone is used. Then,

- Juan's cellular phone costs $\$12.95 + \$0.25x$.
- Tiffany's cellular phone costs $\$14.95 + \$0.15x$.
- The two plans cost the same when

$$\$12.95 + \$0.25x = \$14.95 + \$0.15x$$
$$\$0.25x - \$0.15x = \$14.95 - \$12.95$$
$$\$0.10x = \$2.00$$
$$x = \frac{\$2.00}{\$0.10} = 20$$

The two plans cost the same when each phone is used **20 minutes**.

PART IV

31. The given equation is:

In a proportion, the product of the means is equal to the product of the extremes (cross-multiply):

Rewrite the quadratic equation so that all the nonzero terms are on the same side of the equation and 0 is on the other side:

$$\frac{1}{x} = \frac{x + 1}{6}$$

$$x(x + 1) = (1)(6)$$

$$x^2 + x - 6 = 0$$

Factor the left side of the quadratic equation
as the product of two binomials: $(x + ?) (x + ?) = 0$

The missing terms of the binomials factors are
the two numbers whose sum is +1, the coefficient of
the x-term of $x^2 + x - 6$, and whose product is –6, the
last term of $x^2 + x - 6$. Since $(+3) + (-2) = +1$ and
$(+3)(-2) = -6$, the missing terms are 3 and –2: $(x + 3) (x - 2) = 0$

If the product of two numbers is 0, then at
least one of these numbers is 0: $x + 3 = 0$ or $x - 2 = 0$

$$x = -3 \text{ or } x = 2$$

The solutions are **2** and **–3**.

32. It is given that, on a science quiz, 20 students received the following
scores: 100, 95, 95, 90, 85, 85, 85, 80, 80, 80, 80, 75, 75, 75, 70, 70, 65, 65, 60,
55. Display the data either in a histogram or in a stem-and-leaf plot.

- To construct a histogram, follow these steps:
  STEP 1: Organize the data in a frequency
  table in which the width of each interval
  is the same.
  STEP 2: Label the horizontal axis, "Interval
  of Scores" and the vertical axis, "Number
  of Scores."
  STEP 3: Construct the histogram as shown in

Interval	Number of Scores
91–100	3
81–90	4
71–80	7
61–70	4
51–60	2

the accompanying diagram. Title the histogram, "Science Quiz Scores."

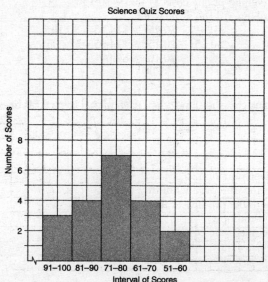

Science Quiz Scores

• Organize the data in a stem-and-leaf plot in which the leaf is the units digit of quiz score and the stem is the remaining part of the score, as shown in the accompanying figure.

Science Quiz Scores

10	0					
9	0	5	5			
8	0	0	0	0	5	5	5
7	0	0	5	5	5	
6	0	5	5			
5	5					

Key: 9 | 5 means a score of 95.

33. In general, the graph of the equation $x^2 + y^2 = r^2$ is a circle whose center is at the origin and whose radius is r. It is given that the equation $x^2 + y^2 = 9$ represents a garden. Hence, the garden is circular with a radius $r = \sqrt{9} = 3$.

a. To graph $x^2 + y^2 = 9$ on the grid that is provided, put the point of your compass on the origin and set the radius length of the compass by placing the pencil point three boxes along the x-axis (or y-axis). Keeping this setting fixed, draw a circle as shown on the accompanying grid.

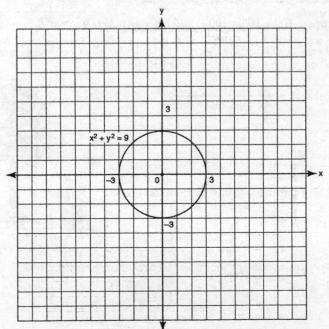

b. The area A of a circle with radius r is given by the formula $A = \pi r^2$. Since $r^2 = 9$, $A = \pi \times 9 \approx 3.14 \times 9 = 28.26$.

The area of the circular garden to the nearest square unit is **28**.

34. a. It is given that 100 more balcony tickets than main-floor tickets were sold for a concert. If x represents the number of main-floor tickets, then $x + 100$ represents the number of balcony tickets.

- Since the balcony tickets sold for \$4, the dollar amount of sales of balcony tickets was $\$4(x + 100)$.
- The main-floor tickets sold for \$12, so the dollar amount of sales of main-floor tickets was \$12x.
- Since the total sales for both types of tickets was \$3,056,

$$4(100 + x) + 12x = 3{,}056$$

b. Solve the equation written in part a:

$$4(100 + x) + 12x = 3{,}056$$
$$400 + 4x + 12x = 3{,}056$$
$$16x = 3{,}056 - 400$$
$$x = \frac{2{,}656}{16} = 166$$

Hence, $x + 100 = 166 + 100 = $ **266** balcony tickets were sold.

35. In the accompanying diagram, the height of the tree is represented by x. To find x, use the tangent ratio in right triangle ABC:

$$\tan A = \frac{\text{leg opposite to } \angle A}{\text{leg adjacent to } \angle A}$$
$$= \frac{BC}{AC}$$
$$\tan 62° = \frac{x}{15}$$

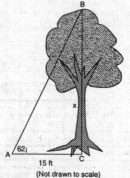

Use a calculator to find that $\tan 62° \approx 1.881$:

$$\frac{x}{15} = 1.881$$
$$x = 1.881 \times 15 = 28.215$$

To the *nearest tenth of a foot*, the height of the tree is **28.2 feet**.

Topic	Question Numbers	Number of Points	Your Points	Your Percentage
1. Numbers; Properties of Real Numbers; Percent	7	2		
2. Operations on Rat'l Numbers & Monomials	—	—		
3. Laws of Exponents for Integer Exponents; Scientific Notation	11	2		
4. Operations on Polynomials	8	2		
5. Square Root; Operations with Radicals	3	2		
6. Evaluating Formulas & Alg. Expressions	13	2		
7. Solving Linear Eqs. & Inequalities	—	—		
8. Solving Literal Eqs. & Formulas for a Particular Letter	16	2		
9. Alg. Operations (including factoring)	5, 9	2 + 2 = 4		
10. Solving Quadratic Eqs., Graphs of Parabolas & Circles	31, 33	4 + 4 = 8		
11. Coordinate Geometry (graphs of linear eqs.; slope; midpoint; distance)	15	2		
12. Systems of Linear Eqs. & Inequalities (alg. and graph. solutions)	—	—		
13. Mathematical Modeling Using: Eqs., Tables, Graphs of Linear Eqs.	21	2		
14. Linear-Quad Systems (alg. and graph. solutions)	—	—		
15. Word Problems Requiring Arith. or Algebraic Reasoning	1, 4, 17, 20, 22, 25, 30	2 + 2 + 2 + 2 + 2 + 2 + 3 = 15		
16. Areas, Perims., Circumf., Vols. of Common Figures	23	2		
17. Angle & Line Relationships (suppl., compl., vertical angles; parallel lines; congruence)	28	3		
18. Ratio & Proportion (incl. similar polygons)	—	—		
19. Pythagorean Theorem	27b	1		
20. Right Triangle Trig. & Indirect Measurement	35	4		
21. Logic (conditionals; logically equiv. statements; valid arguments)	12, 29	2 + 3 = 5		
22. Probability (incl. tree diagrams & sample spaces)	6, 26	2 + 3 = 5		
23. Counting Methods & Sets	14	2		
24. Permutations & Combinations	—	—		
25. Statistics (mean, percentiles, quartiles; freq. dist.; histograms; stem & leaf plots; box-and-whisker plots; circle graphs)	18, 32	2 + 4 = 6		
26. Properties of Triangles & Parallelograms	2, 19	2 + 2 = 4		
27. Transformations (reflections; translations; rotations; dilations)	—	—		

Topic	Question Numbers	Number of Points	Your Points	Your Percentage
28. Symmetry	10	2		
29. Area & Transformations Using Coordinates	24	2		
30. Locus & Constructions	27a	2		
31. Dimensional Analysis & Units of Measurement	—	—		

MAP TO LEARNING STANDARDS

Key Ideas	Item Numbers
Mathematical Reasoning	12, 29
Number and Numeration	7, 22
Operations	2, 3, 5, 8, 9, 10, 11, 20
Modeling/Multiple Representation	1, 16, 19, 24, 27, 28
Measurement	15, 17, 18, 21, 23, 25, 32, 35
Uncertainty	6, 14, 26
Patterns/Functions	4, 13, 30, 31, 33, 34

HOW TO CONVERT YOUR RAW SCORE TO YOUR MATH A REGENTS EXAMINATION SCORE

Below is the conversion chart that must be used to determine your final score on the January 2001 Regents Examination in Math A. To find your final exam score, locate in the column labeled "Raw Score" the total number of points you scored out of a possible 85 points. Since partial credit is allowed in Parts II, III, and IV of the test, you may need to approximate the credit you would receive for a solution that is not completely correct. Then locate in the adjacent column to the right the scaled score that corresponds to your raw score. The scaled score is your final Math A Regents Examination score.

Regents Examination in Math A—January 2001
Chart for Converting Total Test Raw Scores to
Final Examination Scores (Scaled Scores)

Raw Score	Scaled Score	Raw Score	Scaled Score	Raw Score	Scaled Score
85	100	56	75	27	45
84	99	55	74	26	44
83	99	54	73	25	43
82	98	53	72	24	42
81	98	52	71	23	41
80	97	51	70	22	40
79	96	50	69	21	39
78	95	49	68	20	38
77	94	48	67	19	37
76	94	47	66	18	36
75	93	46	65	17	35
74	92	45	64	16	34
73	91	44	63	15	33
72	90	43	62	14	32
71	89	42	61	13	31
70	88	41	60	12	30
69	87	40	59	11	29
68	86	39	58	10	28
67	86	38	57	9	27
66	85	37	56	8	26
65	84	36	55	7	25
64	83	35	53	6	24
63	82	34	52	5	22
62	81	33	51	4	19
61	80	32	50	3	15
60	79	31	49	2	10
59	78	30	48	1	5
58	77	29	47	0	0
57	76	28	46		

Examination
June 2001
Math A

PART I

Answer all questions in this part. Each correct answer will receive 2 credits. No partial credit will be allowed. Record your answers in the spaces provided. [40]

1 A car travels 110 miles in 2 hours. At the same rate of speed, how far will the car travel in h hours?

(1) $55h$ (3) $\dfrac{h}{55}$

(2) $220h$ (4) $\dfrac{h}{220}$ 1 _____

2 Which polynominal is the quotient of $\dfrac{6x^3 + 9x^2 + 3x}{3x}$?

(1) $2x^2 + 3x + 1$ (3) $2x + 3$

(2) $2x^2 + 3x$ (4) $6x^2 + 9x$ 2 _____

3 If the length of a rectangular prism is doubled, its width is tripled, and its height remains the same, what is the volume of the new rectangular prism?

(1) double the original volume
(2) triple the original volume
(3) six times the original volume
(4) nine times the original volume   3 _____

4 One root of the equation $2x^2 - x - 15 = 0$ is

  (1) $\frac{5}{2}$ (3) 3

  (2) $\frac{3}{2}$ (4) –3   4 ____

5 Which properties best describe the coordinate graph of two distinct parallel lines?

  (1) same slopes and same intercepts

  (2) same slopes and different intercepts

  (3) different slopes and same intercepts

  (4) different slopes and different intercepts   5 ____

6 Which statement is *not* always true about a parallelogram?

  (1) The diagonals are congruent.

  (2) The opposite sides are congruent.

  (3) The opposite angles are congruent.

  (4) The opposite sides are parallel.   6 ____

7 In isosceles triangle *DOG*, the measure of the vertex angle is three times the measure of one of the base angles. Which statement about $\triangle DOG$ is true?

  (1) $\triangle DOG$ is a scalene triangle.

  (2) $\triangle DOG$ is an acute triangle.

  (3) $\triangle DOG$ is a right triangle.

  (4) $\triangle DOG$ is an obtuse triangle.   7 ____

8 Which equation illustrates the distributive property for real numbers?

   (1) $\frac{1}{3} + \frac{1}{2} = \frac{1}{2} + \frac{1}{3}$

   (2) $\sqrt{3} + 0 = \sqrt{3}$

   (3) $(1.3 \times 0.07) \times 0.63 = 1.3 \times (0.07 \times 0.63)$

   (4) $-3(5 + 7) = (-3)(5) + (-3)(7)$ 8 _____

9 Factor completely: $3x^2 - 27$

   (1) $3(x - 3)^2$ (3) $3(x + 3)(x - 3)$
   (2) $3(x^2 - 27)$ (4) $(3x + 3)(x - 9)$ 9 _____

10 At a school costume party, seven girls wore masks and nine boys did not. If there were 15 boys at the party and 20 students did not wear masks, what was the total number of students at the party?

   (1) 30   (3) 35
   (2) 33   (4) 42   10 _____

11 If one-half of a number is 8 less than two-thirds of the number, what is the number?

   (1) 24   (3) 48
   (2) 32   (4) 54   11 _____

12 Which statement is logically equivalent to "If I eat, then I live"?

   (1) If I live, then I eat.
   (2) If I eat, then I do not live.
   (3) I live if and only if I eat.
   (4) If I do not live, then I do not eat.   12 _____

13 If a is an odd number, b an even number, and c an odd number, which expression will always be equivalent to an odd number?

(1) $a(bc)$ (3) $ac(b)^1$
(2) $ac(b)^0$ (4) $ac(b)^2$ 13 _____

14 If there are four teams in a league, how many games will have to be played so that each team plays every other team once?

(1) 6   (3) 3
(2) 8   (4) 16   14 _____

15 A woman has a ladder that is 13 feet long. If she sets the base of the ladder on level ground 5 feet from the side of a house, how many feet above the ground will the top of the ladder be when it rests against the house?

(1) 8   (3) 11
(2) 9   (4) 12   15 _____

16 A boy got 50% of the questions on a test correct. If he had 10 questions correct out of the first 12, and $\frac{1}{4}$ of the remaining questions correct, how many questions were on the test?

(1) 16   (3) 26
(2) 24   (4) 28   16 _____

17 A hotel charges \$20 for the use of its dining room and \$2.50 a plate for each dinner. An association gives a dinner and charges \$3 a plate but invites four nonpaying guests. If each person has one plate, how many paying persons must attend for the association to collect the exact amount needed to pay the hotel?

    (1) 60   (3) 40

    (2) 44   (4) 20   17 _____

18 In the set of positive integers, what is the solution set of the inequality $2x - 3 < 5$?

    (1) {0,1,2,3}   (3) {0,1,2,3,4}

    (2) {1,2,3}   (4) {1,2,3,4}   18 _____

19 What is the total number of points of intersection in the graphs of the equations $x^2 + y^2 = 16$ and $y = 4$?

    (1) 1   (3) 3

    (2) 2   (4) 0   19 _____

20 Which is a rational number?

    (1) $\sqrt{8}$ (3) $5\sqrt{9}$

    (2) π (4) $6\sqrt{2}$ 20 _____

PART II

Answer all questions in this part. Each correct answer will receive 2 credits. Clearly indicate the necessary steps, including appropriate formula substitutions, diagrams, graphs, charts, etc. For all questions in this part, a correct numerical answer with no work shown will receive only 1 credit. [10]

21 A school district offers hockey and basketball. The result of a survey of 300 students showed:

> 120 students play hockey, only
> 90 students play basketball, only
> 30 students do not participate in either sport

Of those surveyed, how many students play both hockey and basketball?

22 In the accompanying diagram, parallel lines $\overline{AB}$ and $\overline{CD}$ are intersected by transversal $\overline{EF}$ at points X and Y, and m$\angle FYD$ = 123. Find m$\angle AXY$.

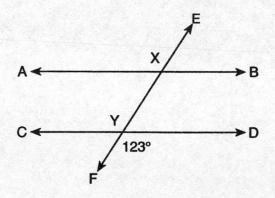

23 Ben had twice as many nickels as dimes. Altogether, Ben had $4.20. How many nickels *and* how many dimes did Ben have?

24 If a girl 1.2 meters tall casts a shadow 2 meters long, how many meters tall is a tree that casts a shadow 75 meters long at the same time?

25 There were seven students running in a race. How many different arrangements of first, second, and third place are possible?

PART III

Answer all questions in this part. Each correct answer will receive 3 credits. Clearly indicate the necessary steps, including appropriate formula substitutions, diagrams, graphs, charts, etc. For all questions in this part, a correct numerical answer with no work shown will receive only 1 credit. [15]

26 In the accompanying diagram of parallelogram $ABCD$, $m\angle A = (2x + 10)$ and $m\angle B = 3x$. Find the number of degrees in $m\angle B$.

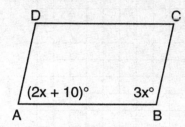

27 A factory packs CD cases into cartons for a music company. Each carton is designed to hold 1,152 CD cases. The Quality Control Unit in the factory expects an error of less than 5% over or under the desired packing number. What is the *least* number and the *most* number of CD cases that could be packed in a carton and still be acceptable to the Quality Control Unit?

28 Connor wants to compare Celsius and Fahrenheit temperatures by drawing a conversion graph. He knows that $-40°C = -40°F$ and that $20°C = 68°F$. On the accompanying grid, construct the conversion graph and, using the graph, determine the Celsius equivalent of $25°F$.

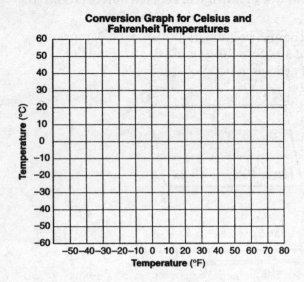

Conversion Graph for Celsius and Fahrenheit Temperatures

29 Virginia has a circular rug on her square living room floor, as represented in the accompanying diagram. If her entire living room floor measures 100 square feet, what is the area of the part of the floor covered by the rug?

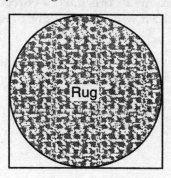

30 Mr. Yee has 10 boys and 15 girls in his mathematics class. If he chooses two students at random to work on the blackboard, what is the probability that both students chosen are girls?

PART IV

Answer all questions in this part. Each correct answer will receive 4 credits. Clearly indicate the necessary steps, including appropriate formula substitutions, diagrams, graphs, charts, etc. For all questions in this part, a correct numerical answer with no work shown will receive only 1 credit. [20]

31 Find three consecutive odd integers such that the product of the first and the second exceeds the third by 8.

32 Keesha wants to tile the floor shown in the accompanying diagram. If each tile measures 1 foot by 1 foot and costs $2.99, what will be the total cost, including an 8% sales tax, for tiling the floor?

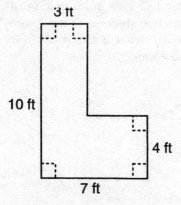

33 Ramón rented a sprayer and a generator. On his first job, he used each piece of equipment for 6 hours at a total cost of $90. On his second job, he used the sprayer for 4 hours and the generator for 8 hours at a total cost of $100. What was the hourly cost of *each* piece of equipment?

34 The plan of a parcel of land is represented by trapezoid *ABCD* in the accompanying diagram. If the area of △*ABE* is 600 square feet, find the minimum number of feet of fence needed to completely enclose the entire parcel of land, *ABCD*.

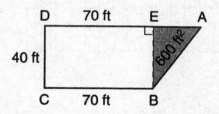

35 Triangle *SUN* has coordinates *S*(0,6), *U*(3,5), and *N*(3,0). On the accompanying grid, draw and label △*SUN*. Then, graph and state the coordinates of △*S′U′N′*, the image of △*SUN* after a reflection in the *y*-axis.

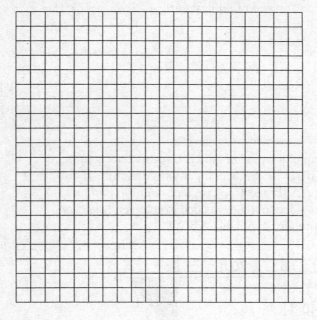

Scrap Graph Paper — This sheet will *not* be scored.

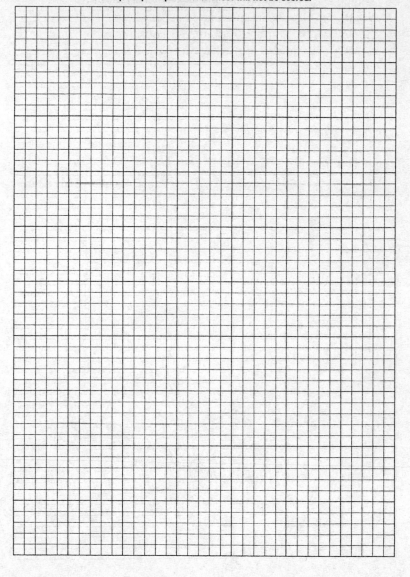

Scrap Graph Paper — This sheet will *not* be scored.

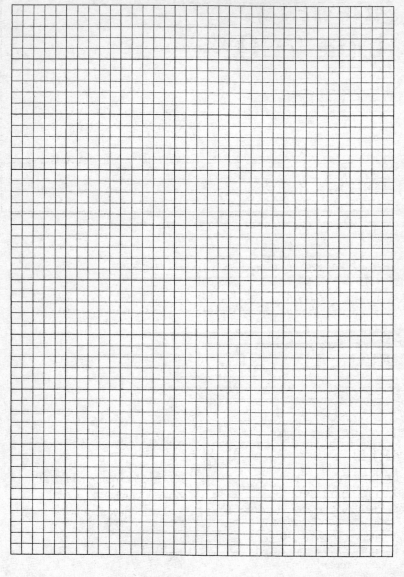

Answers
June 2001
Math A

Answer Key

PART I

1. (1)	**5.** (2)	**9.** (3)	**13.** (2)	**17.** (1)
2. (1)	**6.** (1)	**10.** (2)	**14.** (1)	**18.** (2)
3. (3)	**7.** (4)	**11.** (3)	**15.** (4)	**19.** (1)
4. (3)	**8.** (4)	**12.** (4)	**16.** (4)	**20.** (3)

PART II

21. 60

22. 57°

23. 21 dimes and 42 nickels

24. 45

25. 210

PART III

26. 102

27. 1,095 and 1,209

28. See graph in the *Answers Explained* section; any answer between –6°C and –2°C will be scored as correct.

29. 78.5 or 25π square feet or an equivalent answer

30. $\frac{7}{20}$ or an equivalent answer

PART IV

31. 3, 5, and 7

32. $148.54

33. $5 for the sprayer and
      $10 for the generator

34. 260

35. $S'(0,6)$, $U'(-3,5)$, $N'(-3,0)$.
      See graph in the *Answers
      Explained* section.

Parts II–IV You are required to show how you arrived at your answers. For sample methods of solution, see Answers Explained section.

Answers Explained

PART I

1. Since it is given that a car travels 110 miles in 2 hours, the car travels at an average rate of speed of $\frac{110 \text{ miles}}{2 \text{ hours}} = 55 \frac{\text{miles}}{\text{hour}}$. In h hours, traveling at the same rate of speed, the car will travel $55 \frac{\text{miles}}{\text{hour}} \times h$ hours $= 55h$ miles.

The correct choice is **(1)**.

2. To divide a polynomial by a monomial, divide each term of the polynomial by the monomial:

$$\frac{6x^3 + 9x^2 + 3x}{3x} = \frac{6x^3}{3x^1} + \frac{9x^2}{3x^1} + \frac{3x^1}{3x^1}$$

For each fraction on the right side of the equation divide the numerical coefficient of the numerator by 3 and divide the powers of the same base, x, by subtracting their exponents:

$$= 2x^{3-1} + 9x^{2-1} + 1x^0$$

Let $x^0 = 1$: $\qquad\qquad = 2x^2 + 3x + 1$

The correct choice is **(1)**.

3. The volume of a rectangular prism is the product of its length (ℓ), width (w), and height (h).

- Suppose $\ell = w = h = 1$. Then the volume of the original rectangular prism is $1 \times 1 \times 1 = 1$.
- If its length is doubled, its width is tripled, and its height remains the same, then $\ell = 2, w = 3, h = 1$, and the new volume is $2 \times 3 \times 1 = 6$.
- Hence the volume of the new rectangular prism is six times the original volume.

The correct choice is **(3)**.

4. The given equation, $2x^2 - x - 15 = 0$, is a quadratic equation. To find a root, solve the equation algebraically by factoring, or substitute each of the answer choices into the equation until you find the one that works.

<u>Method 1: Solve algebraically</u>

The quadratic equation $2x^2 - x - 15 = 0$ can be solved by factoring the quadratic trinomial as the product of two binomials:

$$(2x + ?)(x + ?) = 0$$

The missing terms of the binomial factors are the two integers whose product is –15 and that make the sum of the outer and inner products of the binomial factors equal to –1, the coefficient of the x-term of $2x^2 - x - 15$:

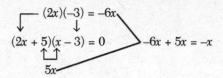

Set each factor equal to 0:

$$2x + 5 = 0 \qquad\qquad x - 3 = 0$$
$$2x = -5 \qquad\qquad x = 3$$
$$x = -\frac{5}{2}$$

The correct choice is **(3)**.

<u>Method 2: Solve by substituting the answer choices</u>
Substitute the answer choices into the quadratic equation beginning with the integers in choice (3) and, if necessary, choice (4):

$$
\begin{array}{ll}
& 2x^2 - x - 15 = 0 \\
\text{Let } x = 3: & 2(3)^2 - 3 - 15 = 0 \\
& 2 \cdot 9 - 18 = 0 \\
& 18 - 18 = 0
\end{array}
$$

5. If two distinct lines are parallel, their slopes are equal. Since parallel lines do not intersect, parallel lines must have different intercepts.

The correct choice is **(2)**.

6. In a parallelogram, opposite sides are parallel, opposite sides are congruent, and opposite angles are congruent. The diagonals of a parallelogram are congruent only in the special cases where the parallelogram is a rectangle or a square.

The correct choice is **(1)**.

7. It is given that, in isosceles triangle *DOG*, the measure of the vertex angle is three times the measure of one of the base angles. If x represents the measure of each base angle, then $3x$ represents the measure of the vertex angle. Since the sum of the measures of the three angles of a triangle is 180:

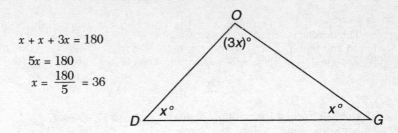

$$x + x + 3x = 180$$
$$5x = 180$$
$$x = \frac{180}{5} = 36$$

Hence the measures of the three angles of $\triangle DOG$ are 36, 36, and $3 \cdot 36 = 108$. Because the measure of one of the angles of $\triangle DOG$ is greater than 90, $\triangle DOG$ is an obtuse triangle.

The correct choice is **(4)**.

8. The distributive property for real numbers states that an expression of the form $a(b + c)$ can be rewritten by multiplying each term inside the parentheses by the term in front of the parentheses, as in $-3(5 + 7) = (-3)(5) + (-3)(7)$.

The correct choice is **(4)**.

9. The given expression is $3x^2 - 27$
        Factor out 3 from each term: $3(x^2 - 9)$
        Since $x^2 - 9$ is the difference of two squares,
it can be factored as the sum and the difference of
the terms that are being squared: $3(x + 3)(x - 3)$

The correct choice is **(3)**.

10. It is given that at a school costume party, 7 girls wore masks, 9 boys did not wear masks, 15 boys were at the party, and 20 students did not wear masks. Organize this information in a table. Since 9 of the 20 students who did not wear masks were boys, $20 - 9 = 11$ girls did not wear masks.

	Boys	Girls
Wore masks	6	7
Did not wear masks	9	$20 - 9 = 11$
Total	15	18

Hence 15 boys + 18 girls = 33 students were at the party.

The correct choice is **(2)**.

11. If one-half of a number $x\left(\dfrac{x}{2}\right)$ is (equals) 8 less than two-thirds of the number $\left(\dfrac{2}{3}x - 8\right)$, then

$$\frac{x}{2} = \frac{2}{3}x - 8$$

Eliminate fractions by multiplying each member of the equation by 6, the lowest common multiple of the denominators:

$$6\left(\frac{x}{2}\right) = 6\left(\frac{2}{3}x\right) - 6 \cdot 8$$
$$3x = 4x - 48$$
$$-x = -48$$
$$x = 48$$

The correct choice is **(3)**.

12. Two statements are logically equivalent when they always have the same truth values. A conditional statement and its contrapositive are logically equivalent statements. To obtain the contrapositive of the conditional statement

If *statement* 1, then *statement* 2,

form the new conditional statement,

If not *statement* 2, then not *statement* 1.

Hence the statement that is logically equivalent to the given statement "If I eat, then I live" is its contrapositive, "If I do not live, then I do not eat."

The correct choice is **(4)**.

13. It is given that a is an odd number, b is an even number, and c is an odd number. Pick easy numbers for a, b, and c. Let $a = 1$, $b = 2$, and $c = 3$. Evaluate the expressions in each of the answer choices:

- Choice (1): $a(bc) = 1(2 \cdot 3) = 6$. This is not the correct answer choice.
- Choice (2): $ac(b)^0 = 1 \cdot 3 \cdot 1$ because any nonzero number raised to the zero power is 1. Since $ac(b)^0 = 3$, this is the correct answer choice.
- Choice (3): $ac(b)^1 = 1 \cdot 3(2)^1 = 3 \cdot 2 = 6$. This is not the correct answer choice.
- Choice (4): $ac(b)^2 = 1 \cdot 3(2)^2 = 3 \cdot 4 = 12$. This is not the correct answer choice.

The correct choice is **(2)**.

14. It is given that there are four teams in a league. To find the number of games that will have to be played so that each team plays every other team once, use combinations or make an organized list.

<u>Method 1: Use combinations</u>

The number of games that will have to be played is the number of combinations of the four teams taken two at a time, which is represented by the expression $_4C_2$. To evaluate $_4C_2$, use a scientific calculator or the formula

$$_nC_r = \frac{n!}{r!(n-r)!}$$

where $n = 4$ and $r = 2$:

$$_4C_2 = \frac{4!}{2!(4-2)!}$$

$$= \frac{\overset{2}{4}\times 3\times\cancel{2\times 1}}{(\cancel{2\times 1})(\overset{1}{\cancel{2\times 1}})}$$

$$= 6$$

The correct choice is **(1)**.

<u>Method 2: Make a list</u>

Call the teams A, B, C, and D. List all possible combinations or pairings of the two teams without regard to order:

1. AB
2. AC
3. AD
4. BC
5. BD
6. CD

15. It is given that a woman sets the base of a ladder 13 feet in length on level ground 5 feet from the side of a house. To find the number of feet above the ground the top of the ladder will be when it rests against the house, find the height of the right triangle whose hypotenuse is the length of the ladder:

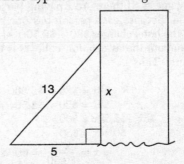

$$x^2 + 5^2 = 13^2$$
$$x^2 + 25 = 169$$
$$x^2 = 169 - 25 = 144$$
$$x = \sqrt{144} = 12$$

The correct choice is **(4)**.

16. It is given that a boy got 50% $\left(=\frac{1}{2}\right)$ of the questions on a test correct. It is also given that he had 10 questions correct out of the first 12, and $\frac{1}{4}$ of the remaining questions correct. If x represents the total number of questions on the test, then the number of questions the boy had correct was 10 plus $\frac{1}{4}$ of $x - 12$. Thus

$$\frac{\text{Number of correct questions}}{\text{Total number of questions}} = \frac{10 + \frac{1}{4}(x - 12)}{x} = \frac{1}{2}$$

Cross-multiply:

$$2\left[10 + \frac{1}{4}(x - 12)\right] = x$$
$$20 + \frac{1}{2}(x - 12) = x$$
$$20 + \frac{1}{2}x - 6 = x$$
$$14 = x - \frac{1}{2}x$$
$$14 = \frac{1}{2}x$$
$$2 \cdot 14 = x$$
$$28 = x$$

The correct choice is **(4)**.

17. It is given that
- A hotel charges $20 for the use of its dining room and $2.50 a plate for each dinner.
- An association gives a dinner and charges $3 a plate but invites four non-paying guests. If there are x paying persons, the total number of persons is $x + 4$. Because each person has one plate, the association collects $3x$ while the hotel charges $20 + $2.50(x + 4)$.
- The amount the association collects is the exact amount needed to pay the hotel. Thus

$$\$3x = \$20 + \$2.50(x + 4)$$
$$\$3x = \$20 + \$2.50x + \$10.00$$
$$\$3x - \$2.50x = \$30$$
$$\$0.50x = \$30$$
$$x = \frac{\$30}{\$0.50} = 60$$

The correct choice is **(1)**.

18. If $2x - 3 < 5$, then $2x < 8$, so $x < \frac{8}{2}$, or $x < 4$. If x represents a positive integer, then $x < 4$ represents the set of positive integers less than 4, which is $\{1,2,3\}$. The number 0 is not included in the solution set since 0 is neither positive nor negative.

The correct choice is **(2)**.

19. The graph of an equation that has the form $x^2 + y^2 = r^2$ is a circle with radius r whose center is at the origin. Hence, the graph of $x^2 + y^2 = 16$ is a circle with radius $\sqrt{16} = 4$ whose center is at the origin, as shown in the accompanying figure. The graph of $y - 4$ is a horizontal line 4 units above the x-axis. Hence the graphs of $x^2 + y^2 = 16$ and $y = 4$ intersect at one point at $(0,4)$.

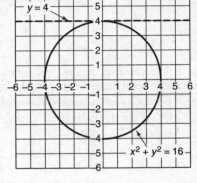

The correct choice is **(1)**.

20. A rational number is any number that can be written as an integer or as the quotient of two integers. Find the answer choice containing an expression that can be written in this form. Choice (3) contains a rational number since $5\sqrt{9} = 5 \cdot 3 = 15$.

The correct choice is **(3)**.

PART II

21. It is given that the results of a survey of 300 students show that

    120 students play only hockey
    90 students play only basketball
    30 students do not participate in either sport

This represents a total of 240 students. The only possibility that has not been accounted for is the number of students who play both hockey and basketball. Since 300 students were surveyed, $300 - 240 = 60$ students play both hockey and basketball.

60 students play both hockey and basketball.

22.

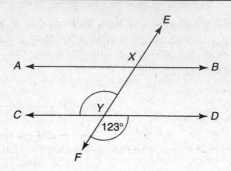

It is given that, in the accompanying diagram, transversal $\overline{EF}$ intersects parallel lines AB and CD at points X and Y, and m$\angle FYD = 123$.

- Because vertical angles are equal in measure, m$\angle CYE = 123$.
- Since $\overleftrightarrow{AB} \parallel \overleftrightarrow{CD}$, interior angles on the same side of the transversal are supplementary. Hence

$$m\angle AXY = 180 - m\angle CYE$$
$$= 180 - 123$$
$$= 57$$

The measure of $\angle AXY$ is **57**.

23. It is given that Ben had twice as many nickels as dimes. If Ben had x dimes, he had $2x$ nickels. It is also given that Ben had \$4.20. Hence

$$\$0.10x + \$0.05(2x) = \$4.20$$
$$\$0.10x + \$0.10x = \$4.20$$
$$\$0.20x = \$4.20$$
$$x = \frac{\$4.20}{\$0.20} = 21$$

Hence Ben had **21** dimes and $2(21) = $ **42** nickels.

24. If a girl 1.2 meters tall casts a shadow 2 meters long and a tree x meters tall casts a shadow 75 meters long, then

$$\frac{\text{Height}}{\text{Shadow}} = \frac{1.2}{2} = \frac{x}{75}$$
$$2x = 75(1.2)$$
$$x = \frac{90}{2} = 45$$

The tree is **45 meters** tall.

25. If 7 students are running in a race, then any one of the 7 students can be in the first-place position, any one of the remaining 6 students can fill the second-place position, and any one of the remaining 5 students can fill the third-place position. The total number of different ordered possible arrangements of first, second, and third places is given by the product of 7, 6, and 5:

1st place		2nd place		3rd place		
$\boxed{7}$	$\times$	$\boxed{6}$	$\times$	$\boxed{5}$	$=$	210

There are **210** possible arrangements of first, second, and third places.

PART III

26. It is given that, in the accompanying diagram of parallelogram $ABCD$, $m\angle A = 2x + 10$ and $m\angle B = 3x$. Since consecutive angles of a parallelogram are supplementary,

$$m\angle A + m\angle B = 180$$
$$(2x + 10) + 3x = 180$$
$$5x = 180 - 10$$
$$x = \frac{170}{5} = 34$$

Hence $m\angle B = 3x = 3(34) = 102$.

27. It is given that a carton is designed to hold 1,152 CD cases and that the Quality Control Unit expects an error of less than 5% of the desired packing number of 1,152.

- Since 5% of $1{,}152 = 0.05 \times 1{,}152 = 57.6$, round down to 57 to find the number of packed cases that is *within* 5% under and 5% above 1,152.
- The *least* number of CD cases that could be packed in a carton and still be acceptable to the Quality Control Unit is $1{,}152 - 57 = \mathbf{1{,}095}$.
- The *most* number of CD cases that could be packed in a carton and still be acceptable to the Quality Control Unit is $1{,}152 + 57 = \mathbf{1{,}209}$.

28.

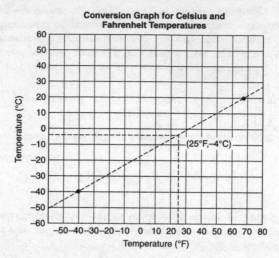

To construct the graph for converting between degrees Fahrenheit and degrees Celsius, plot the given points (–40°F,–40°C) and (–68°F,–20°C) on the grid that is provided. Then use a straightedge to draw a line through these points. From the graph, estimate that when the x-coordinate of a point on the line is 25°F, the corresponding y-coordinate is approximately –4°C. Any answer between –6°C and –2°C will be scored as correct.

29.

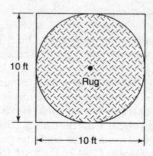

It is given that, in the accompanying diagram, the area of the square living room is 100 square feet.

- Because $10 \times 10 = 100$, each side of the square measures 10 feet.
- The diameter of the circular rug is equal to the length of a side of the square. Hence the radius of the circular rug is $\frac{1}{2} \times 10 = 5$ feet.
- The area of the circular rug $= \pi \times (radius)^2 = \pi \times (5)^2 = 25\pi \approx 25(3.14) = 78.5$ square feet.

The number of square feet covered by the rug is 25π or **78.5**.

30. It is given that Mr. Yee has 10 boys and 15 girls in his mathematics class. Mr. Yee chooses two students at random to work at the blackboard.

- Since 15 of the 10 + 15 = 25 students are girls, the probability that the first student chosen will be a girl is $\frac{15}{25}$.

- Since 14 of the remaining 24 students are girls, the probability that the second student chosen will be a girl is $\frac{14}{24}$.

- Hence the probability that he will choose two girls to work at the blackboard is $\frac{15}{25} \times \frac{14}{24} = \frac{3}{5} \times \frac{7}{12} = \frac{21}{60}$ or $\frac{7}{20}$.

PART IV

31. Since consecutive odd (or consecutive even) integers differ by 2, a set of three consecutive odd integers can be represented by x, $x + 2$, and $x + 4$.

It is given that the product of the first and second of two consecutive odd integers $[x(x + 2)]$ exceeds the third $(x + 4)$ by 8. Thus

$$x(x + 2) = (x + 4) + 8$$
$$x^2 + 2x = x + 12$$
$$x^2 + 2x - x - 12 = 0$$
$$x^2 + x - 12 = 0$$

Factor the left side of the quadratic equation as the product of two binomials: $(x + ?)(x + ?) = 0$

The missing terms of the binomial factors are the two integers whose product is -12, the last term of $x^2 + x - 12$, and whose sum is $+1$, the coefficient of the x-term of $x^2 + x - 12$. Since $(-3)(+4) = -12$ and $(-3) + (+4) = -1$, the missing integers in the binomial factors are $+4$ and -3: $(x - 3)(x + 4) = 0$

Set each factor equal to 0: $x - 3 = 0 \quad x + 4 = 0$
$$x = 3 \qquad x = -4$$

Reject the solution $x = -4$ since -4 is not an odd integer. Hence, $x = 3$, $x + 2 = 5$, and $x + 4 = 7$.

The three consecutive odd integers are **3**, **5**, and **7**.

32.

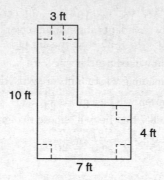

It is given that Keesha wants to tile the floor shown in the accompanying figure using tiles measuring 1 foot by 1 foot each and costing $2.99 each.

Divide the figure into two rectangles, as shown in the accompanying figure. Then find the sum of the areas of the two rectangles:

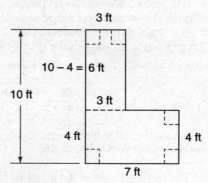

Area of first rectangle   = 3 × 6 = 18 square feet
Area of second rectangle = 7 × 4 = 28 square feet

Total area = 46 square feet

- Find the cost of tiling the floor:

    46 square feet × $2.99 per square foot = $137.54

- Add the 8% sales tax to find the total cost:

    Total cost = $137.54 + 0.08 × $137.54
            = $137.54 + $11.00
            = **$148.54**

33. It is given that Ramón rented a sprayer and a generator.
- On his first job, he used each piece of equipment 6 hours at a total cost of $90. If x represents the hourly cost of the sprayer and y represents the hourly cost of the generator, then $6x + 6y = 90$.
- On his second job, he used the sprayer for 4 hours and the generator for 8 hours at a total cost of $100. Hence $4x + 8y = 100$.
- Solve the system of two equations simultaneously:

$$4x + 8y = 100 \quad \rightarrow \quad \frac{4x}{4} + \frac{8y}{4} = \frac{100}{4} \quad \rightarrow \quad \underline{x + 2y = 25} \qquad \text{Equation (1)}$$

$$6x + 6y = 90 \quad \rightarrow \quad \frac{6x}{6} + \frac{6y}{6} = \frac{90}{6} \quad \rightarrow \quad \underline{x + y = 15} \qquad \text{Equation (2)}$$

Subtract the equations: $\quad 0 + y = 10$

To find x, substitute 10 for y in either of the original equations:

$$4x + 8(10) = 100$$
$$4x = 100 - 80$$
$$x = \frac{20}{4} = 5$$

Thus the hourly cost of the sprayer was **$5** and the hourly cost of the generator was **$10**.

34.

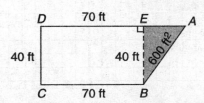

It is given that the accompanying diagram of trapezoid $ABCD$ represents the plan of a parcel of land. To find the minimum number of feet of fence needed to completely enclose the entire parcel of land, calculate the perimeter of trapezoid $ABCD$.
- Since $CBED$ is a rectangle, $BE = CD = 40$. It is also given that the area of $\triangle ABE$ is 600 square feet. Hence

$$\text{Area } \triangle ABE = \frac{1}{2} \times \text{base} \times \text{height}$$
$$600 = \frac{1}{2} \times AE \times BE$$
$$600 = \frac{1}{2} \times AE \times 40$$
$$600 = 20AE$$
$$\frac{600}{20} = AE$$
$$30 = AE$$

- In right triangle ABE, leg $AE = 30$ and leg $BE = 40$. Since the lengths of right triangle ABE form a 3-4-5 Pythagorean triple in which the length of each side is multiplied by 10, the length of hypotenuse AB is $5 \times 10 = 50$.
- $BC + CD + ED + AE + AB = 70 + 40 + 70 + 30 + 50 = 260$ feet.

Thus **260 feet** of fence are needed.

35.

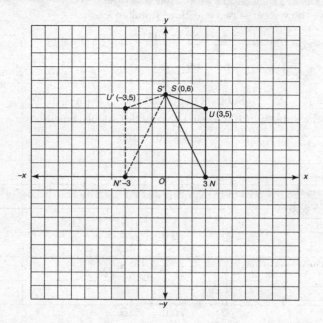

On the grid provided, draw and label $\triangle SUN$. To reflect a point (x,y) in the y-axis, "flip" it over the y-axis so that $(x,y) \rightarrow (-x,y)$. To reflect $\triangle SUN$ in the y-axis, locate the images of each of its vertices:

$$S(0,6) \rightarrow S'(0,6)$$
$$U(3,5) \rightarrow U'(-3,5)$$
$$N(3,0) \rightarrow N'(-3,0)$$

Plot S', U', and N'. Then connect these points to form $\triangle S'U'N'$, as shown in the accompanying figure.

Topic	Question Numbers	Number of Points	Your Points	Your Percentage
1. Numbers; Properties of Real Numbers; Percent	8, 13, 20	2 + 2 + 2 = 6		
2. Operations on Rat'l. Numbers & Monomials	—	—		
3. Laws of Exponents for Integer Exponents; Scientific Notation	—	—		
4. Operations on Polynomials	2	2		
5. Square Root; Operations with Radicals	—	—		
6. Evaluating Formulas & Algebraic Expressions	—	—		
7. Solving Linear Eqs. & Inequalities	18	2		
8. Solving Literal Eqs. & Formulas for a Particular Letter	—	—		
9. Alg. Operations (including factoring)	9	2		
10. Solving Quadratic Eqs.; Graphs of Parabolas & Circles	4	2		
11. Coordinate Geometry (graphs of linear eqs.; slope; midpoint; distance)	5	2		
12. Systems of Linear Eqs. & Inequalities (alg. and graph. solutions)	33	4		
13. Mathematical Modeling Using: Eqs., Tables, Graphs of Linear Eqs.	28, 31	3 + 4 = 7		
14. Linear-Quad Systems (alg. and graph. solutions)	19	2		
15. Word Problems Requiring Arith. or Algebraic Reasoning	1, 11, 16, 17, 23, 27	2 + 2 + 2 + 2 + 2 + 3 = 13		
16. Areas, Perims., Circums., Vols. of Common Figures	3, 29, 32, 34	2 + 3 + 4 + 4 = 13		
17. Angle & Line Relationships (suppl., compl., vertical angles; parallel lines; congruence)	22	2		
18. Ratio & Proportion (incl. similar polygons)	24	2		
19. Pythagorean Theorem	15	2		
20. Right Triangle Trig. & Indirect Measurement	—	—		
21. Logic (symbolic rep.; conditionals; logically equiv. statements; valid arguments)	10. 12, 21	2 + 2 + 2 = 6		
22. Probability (incl. tree diagrams & sample spaces)	30	3		
23. Counting Methods & Sets	—	—		
24. Permutations & Combinations	14, 25	2 + 2 = 4		
25. Statistics (mean, percentiles, quartiles; freq. dist., histograms, stem and leaf plots; box-and-whisker plots; circle graphs)	—	—		
26. Properties of Triangles & Parallelograms	6, 7, 26	2 + 2 + 3 = 7		
27. Transformations (reflections, translations, rotations, dilations)	—	—		

Topic	Question Numbers	Number of Points	Your Points	Your Percentage
28. Symmetry	—	—		
29. Area & Transformations Using Coordinates	34	4		
30. Locus & Constructions	—	—		
31. Dimensional Analysis & Units of Measurement	—	—		

MAP TO LEARNING STANDARDS

Key Ideas	Item Numbers
Mathematical Reasoning	10, 12, 21
Number and Numeration	8, 13, 20
Operations	2, 9, 11, 35
Modeling/Multiple Representation	3, 6, 7, 17, 22, 23, 26, 29, 31
Measurement	1, 5, 15, 16, 24, 27, 32, 34
Uncertainty	14, 25, 30
Patterns/Functions	4, 18, 19, 28, 33

HOW TO CONVERT YOUR RAW SCORE TO YOUR MATH A REGENTS EXAMINATION SCORE

Below is the conversion chart that must be used to determine your final score on the June 2001 Regents Examination in Math A. To find your final exam score, locate in the column labeled "Raw Score" the total number of points you scored out of a possible 85 points. Since partial credit is allowed in Parts II, III, and IV of the test, you may need to approximate the credit you would receive for a solution that is not completely correct. Then locate in the adjacent column to the right the scaled score that corresponds to your raw score. The scaled score is your final Math A Regents Examination score.

Regents Examination in Math A—June 2001
Chart for Converting Total Test Raw Scores to
Final Examination Scores (Scaled Scores)

Raw Score	Scaled Score	Raw Score	Scaled Score	Raw Score	Scaled Score
85	100	56	77	27	39
84	99	55	76	26	38
83	99	54	75	25	36
82	99	53	73	24	35
81	98	52	72	23	34
80	97	51	71	22	32
79	97	50	70	21	31
78	96	49	69	20	29
77	96	48	67	19	28
76	95	47	66	18	26
75	94	46	65	17	25
74	93	45	64	16	23
73	93	44	62	15	22
72	92	43	61	14	20
71	91	42	60	13	19
70	90	41	59	12	18
69	90	40	57	11	16
68	89	39	56	10	15
67	88	38	55	9	13
66	87	37	53	8	12
65	86	36	52	7	10
64	85	35	51	6	9
63	84	34	49	5	7
62	83	33	48	4	6
61	82	32	46	3	4
60	81	31	45	2	3
59	80	30	44	1	1
58	79	29	42	0	0
57	78	28	41		

NOTES